Gabrielle Falloppia, 1522/23–15(

Renaissance anatomist Gabrielle Falloppia is best known today for his account of the eponymous fallopian tubes but he made numerous other anatomical discoveries as well, was one of the most famous surgeons of his time, and is widely believed to have invented the condom.

Drawing on Falloppia's *Observationes anatomicae* of 1561 and on dozens of handwritten and published sets of student notes, this book not only looks at Falloppia's anatomical lectures and demonstrations. It also studies Falloppia's work on surgical topics – including the French disease and cosmetic surgery – on thermal waters, and on pharmacology. Last but not least, it uses student notes and the letters of contemporary scholars to throw a new light on Falloppia's biography, on his very special relationship with the botanist Melchior Wieland, who lived in his house for several years, and on his conflicts with his fellow professors in Padua, one of whom, Bassiano Landi, was murdered just ten days after his funeral – by Falloppia's disciples, as some believed.

Written by one of the leading scholars in the field of early modern medicine, this book will appeal to all those interested in the teaching and practice of anatomy, surgery, and pharmacology in the Renaissance.

Michael Stolberg was originally trained as a physician. In 1994, he received a PhD in history and philosophy at the University of Munich. From 1995, he held fellowships in Venice, Cambridge, and Munich. Since 2004 he has been chair of the history of medicine at the University of Würzburg, Germany. He has published widely on learned medical theory and practice, the patient experience, and body history in early modern Europe.

The History of Medicine in Context

Series Editors: Andrew Cunningham (Department of History and Philosophy of Science, University of Cambridge) and Ole Peter Grell (Department of History, Open University)

Titles in the series include

Civic Medicine
Physician, Polity, and Pen in Early Modern Europe
Edited by J. Andrew Mendelsohn, Annemarie Kinzelbach, and Ruth Schilling

Authority, Gender, and Midwifery in Early Modern Italy
Contested Deliveries
Jennifer F. Kosmin

Forty Days
Quarantine and the Traveller, c. 1700–c. 1900
John Booker

The World of Worm
Physician, Professor, Antiquarian, and Collector, 1588–1654
Ole Peter Grell

'I Follow Aristotle'
How William Harvey Discovered the Circulation of the Blood
Andrew Cunningham

Gabrielle Falloppia, 1522/23–1562
The Life and Work of a Renaissance Anatomist
Michael Stolberg

For more information about this series, please visit: https://www.routledge.com/The-History-of-Medicine-in-Context/book-series/HMC

Gabrielle Falloppia, 1522/23–1562

The Life and Work of a
Renaissance Anatomist

Michael Stolberg

Routledge
Taylor & Francis Group

LONDON AND NEW YORK

Cover image: The branching off of the humeral vein, "according
to Vesalius" (below) and "according to truth" (above), Staats- und
Universitätsbibliothek Göttingen, Ms. Meibom 20, fol. 133r.

First published 2023
by Routledge
4 Park Square, Milton Park, Abingdon, Oxon OX14 4RN

and by Routledge
605 Third Avenue, New York, NY 10158

Routledge is an imprint of the Taylor & Francis Group, an informa business

© 2023 Michael Stolberg

The right of Michael Stolberg to be identified as author of this work
has been asserted in accordance with sections 77 and 78 of the
Copyright, Designs and Patents Act 1988.

British Library Cataloguing-in-Publication Data
A catalogue record for this book is available from the British Library

Library of Congress Cataloging-in-Publication Data
A catalog record has been requested for this book

ISBN: 978-1-032-14970-7 (hbk)
ISBN: 978-1-032-14971-4 (pbk)
ISBN: 978-1-003-24200-0 (ebk)

DOI: 10.4324/9781003242000

Typeset in Times New Roman
by codeMantra

Contents

Introduction

He was one of the most famous anatomists of all times. In his days, he was compared to Herophilos, the great anatomical authority of antiquity. According to some of his contemporaries, he even surpassed the "divine" Vesalius. In his major anatomical work, the *Observationes anatomicae*, he presented a large number of new anatomical findings and he corrected many an error Galen and Vesalius had made. To this day, the ovarian tubes, of which he gave the first accurate description, are known as "fallopian tubes". Among his students were leading anatomists and surgeons of the following generation, men like Volcher Coiter and Girolamo Fabrizi d'Acquapendente. His name also found its way into the history books in a completely different context: he is widely regarded as the inventor of the condom, although, as we will see, this claim has to be taken with a grain of salt.

Considering his undisputed historical importance, modern historical scholarship on Falloppia is decidedly scanty. Even in major recent studies on the history of Renaissance anatomy, Falloppia is mentioned in passing only.[1] To this day, there is only one major monograph on Falloppia, published by Giuseppe Favaro in 1928.[2] Favaro painstakingly collected biographical evidence from the older literature and conducted in-depth archival studies. In an extensive appendix, he published various letters and documents that provide important information about Falloppia's life. Notwithstanding occasional errors and the somewhat eulogical style typical of the period, Favaro's study is of great value to this day. It is primarily a biography, however. As Montalenti remarked at the time of its publication, a comprehensive and systematic analysis of Falloppia's scientific work and his teaching in other areas remained an urgent desideratum.[3] The situation has not changed much since then. In 1970, Pericle di Pietro curated a complete edition of Falloppia's – rather limited – surviving correspondence, assembling and reediting the letters that Giovanni Fantuzzi, Giuseppe Campori, Alfonso Corradi, Alberto Angelini, Giovanni Battista de Toni, Giuseppe Favaro, and others[4] had previously published, and adding a handful of letters that he himself had found.[5] In 1964, Gabriella Righi Riva and Pericle Di Pietro complemented a reprint of Falloppia's *Observationes* with an Italian translation.[6] The best more recent

DOI: 10.4324/9781003242000-1

biographical overviews are Di Pietro's *Contributo alla biografia di Gabriele Falloppia* (1974) and the entry by Gabriella Belloni Speciale in the *Dizionario biografico degli italiani* (1994).[7]

Falloppia's name, I should hasten to add, can be found in countless smaller publications on the history of anatomy, in medical history textbooks, and in biographical encyclopedias of famous scientists and physicians. By presenting Falloppia's findings and discoveries on specific anatomical structures in the light of modern anatomical knowledge, some of these contributions have the merit of making his discoveries more accessible to readers who cannot read Falloppia in the original Latin or even provide an English translation of the relevant passages. Along these lines, Charles D. O'Malley, Pietro Franceschini, and others have highlighted Falloppia's findings on the ovarian tubes, the auditory ossicles, the cranial nerves, and the eye muscles, among others. When complex anatomical structures are concerned, the anatomical training and expertise of the authors sometimes proves helpful as well.[8]

Many articles on Falloppia only offer a summary of what is known already, however, without adding any findings of their own. In extreme cases, eight authors join forces to present nothing but a rehash of the current state of knowledge and do not even bother to quote Falloppia's work.[9] Even worse, many of these publications ignore the current state of knowledge and reproduce the errors, inaccuracies, and inconsistencies of previous writers, especially regarding Falloppia's biography. Not only the year he obtained his doctoral degree, the time periods during which he taught at different universities, the chairs he held, and even the date of his death are often incorrect.[10] To this day, some writers claim – against the unequivocal historical evidence – that Falloppia studied with Vesalius (rather than, as he himself explained, just reading his work)[11] and that he was, like Vesalius, accused of vivisection.[12] Even standard biographical dictionaries cannot always be trusted,[13] and the entries on Falloppia in *Wikipedia* reproduce similar mistakes.[14]

Historical research on Falloppia's work has understandably focused above all on his anatomical findings.[15] This is where he excelled and where many of his findings and achievements are still recognized today. Anatomy was only one of the fields, however, in which Falloppia was active. The majority of his lectures were devoted not to anatomy, in fact, but to various fields and aspects of surgery and he also lectured on simples, thermal springs, and other topics of pharmacology. These activities have so far attracted much less attention than Falloppia's anatomical work.[16] Historians have shown even less interest in Falloppia as a medical practitioner.

This brief sketch of the state of the field already indicates the central aims of this book. Almost a hundred years after Favaro's book and, by a welcome coincidence, on the occasion of the quincentennial of Falloppia's birth, this is the first comprehensive monograph on Falloppia. With regard to his biography, I will add few details to Favaro's account, among others regarding the conflicts with his higher-ranking colleagues in Padua and the murder of his opponent Bassiano Landi just a few days after Falloppia's funeral. My main goal here will be to separate the wheat from the chaff, that is, to point out and correct the numerous errors and inaccuracies that we find in many biographical accounts, old and recent, and to make clear what we actually know and where we can only conjecture. The bulk of this book will be devoted to Falloppia's research, teaching, and medical practice. Drawing on manuscript student notes, I will date some of his anatomical discoveries to years before the publication of the *Observationes*, and I will not only describe his major findings but also his anatomical lectures and demonstrations. Much more than his published work, unpublished student notes also provide evidence for Falloppia's interest in comparative anatomy. Moreover, I will look in considerable detail at his work in the various other fields in which he was active and in which he enjoyed a considerable renown in his days: from surgery and the French disease to thermal springs and medicinal plants. I will do so with a special eye for original or innovative contributions, such as his endorsement of a "palliative" cure, as he already called it, his discussion of medical cosmetics, and his chemical analysis of mineral waters. Of course I will also tell the story of the "condom", the piece of fabric he recommended putting over the glans as a means to prevent an infection with the French disease. And I will look at the experiments on and vivisections of men who had been condemned to capital punishment which historians have attributed to Falloppia – wrongly, as it turns out – while offering substantial new evidence that he did kill a number of these men with deadly doses of opium in order to dissect them afterward. At the end of the book, the reader will find a bibliography of the quite numerous editions of Falloppia's works or more precisely of the various individual and collective publications of student notes on his lectures.

Sources

The rather scant sources on Falloppia's biography have been presented and exploited again and again, from Tiraboschi's detailed entry in his collection of biographies of Modenese scholars of 1782 and Vicenzo Calderato's doctoral thesis on Falloppia (1862) to Favaro's comprehensive biographical study of 1928, and Di Pietro's work on his correspondence.[17] Unfortunately, considerable gaps remain to this day. Especially Falloppia's early years and his turning toward medicine remain largely in the dark.

Among the printed sources, the *Observationes anatomicae* of 1561 takes a prominent place, the only work Falloppia published himself, during his lifetime. It must not be confused with the notes on a handful of anatomical structures Falloppia's demonstrated to his students which Franciscus Michinus published in 1570 under the title *Observationes anathomicae* [sic!] and which were later also reprinted as *Observationes de venis*.[18] For Falloppia's anatomical work and teaching, I will moreover draw to a considerable extent on the detailed notes of two students who attended

Falloppia's anatomical demonstrations in the early 1550s already.[19] One of them filled about approximately 200 pages with notes on Falloppia's anatomical lectures and demonstrations from 1551 to 1553.[20] The manuscript has come down as part of the extensive *nachlass* of the Meiboms, a dynasty of physicians who were active in Helmstedt.[21] In the following, I will therefore refer to the unidentified scribe as "Helmstedt Anonymus". In addition, I will draw on the notes of Georg Handsch, a medical student from Leipa in Bohemia. Handsch had made extensive notes on a public anatomy presided over by Antonio Fracanzano in Padua in the winter of 1550–1551[22] and added what he learned in the following two years from Falloppia in numerous annotations in the margin and on slips of paper which he inserted between the pages of his original manuscript.[23] Some further insights are offered by *De humani corporis compendium* which the brothers Meietus in Venice published in 1571 under Falloppia's name (later reprinted also as *Institutiones anatomicae*), which was, by all appearances, also based on student notes on lectures which preceded the writing of the *Observationes anatomicae*.[24]

Our knowledge of Falloppia's teaching on the broad range of other topics in medicine and natural history he covered in his lectures comes almost exclusively from student notes. Notes on most of his lectures were eventually published by his disciples or anonymously by publishers who got hold of them. Some of these lectures were commentaries on authoritative writings such as Galen's *De ossibus*, the *Materia medica* of Dioscorides, and the Hippocratic *De vulneribus capitis*. Others were topical lectures, such as ulcers and swellings (*De ulceribus, De tumoribus praeter naturam*), injuries and dislocations (*De vulneribus, De fracturis, De luxationibus*), and the French disease (*De morbo gallico*). Falloppia's lectures on thermal springs and on minerals and metals (*De aquis thermalibus, De fossilibus*) devoted considerable time to questions of natural philosophy but ultimately the importance for the physician also lay in the potential therapeutic effects.

Except for a few isolated passages – especially the one on the "condom" – historians have so far paid little attention to these texts. A major reason for their reluctance to engage with them is undoubtedly that they are not authentic works from Falloppia's pen.[25] Falloppia did not publish on these topics and – although this has sometimes been done – the lecture notes cannot be cited simply as if the text had been written by Falloppia himself. Closer analysis and a comparison of the notes of different students on the same lecture leave no doubt, however, that the student-scribes sought to render Falloppia's words as faithfully as they could. In keeping with the format of an oral lecture, the language usually is less complex or convoluted than that of many printed publications. In numerous places, Falloppia's opinions and observations are rendered in the first person, with phrases such as "I have experienced" ("ego expertus sum") and "but I say" ("ego autem dico"). Even the "domini" with which Falloppia sometimes addressed his listeners tends to be retained. Those who put their notes into print did not have any plausible reason either for willfully distorting Falloppia's words. Quite to the contrary: some of their readers would very likely have been fellow students who had attended the same lecture and possibly even kept their own notes. Petrus Angelus Agathus, in his published notes on Falloppia's lectures, even carefully marked his own additions with brackets.[26]

While notes of different students on Falloppia's lectures on the same topic usually agree very well in content, the specific wording sometimes differs more markedly. Some of the students who later published their notes admitted that in the hurry they might not have been able to take down everything word for word and might have omitted or added a word or two.[27] The discrepancies can only in part be attributed to a lack of precision on the part of the student-scribes, however. They were bound to result almost inevitably when students attended Falloppia's lectures on one and the same topic at different times. Accordingly, Bruno Seidel explained the differences between his edition of *De ulceribus* (1577) and the earlier Venetian edition with the fact that the professors, as those who were familiar with teaching in Italy knew, repeated their lectures on one and the same topic about every three years. For this reason, the Venetian edition offered some passages where Falloppia had expressed a different opinion on certain things which Seidel had not wanted to add. At the same time, his edition was much more complete because it also included Falloppia's teaching on malignant and deep ulcers, ulcers of the nose, palate, and buttocks, herpes, cancerous ulcers, fistulas, and related ailments.[28]

As Seidel's remarks already suggest, the printed student notes on the lectures Falloppia gave on one and the same topic sometimes also lacked certain parts or remained a mere fragment. In some years, Falloppia did not get as far as planned with his lecture on a certain topic. The student-scribes, in turn, may also quite simply have missed individual lectures or were altogether absent from Padua for parts of the academic year. For example, the 1563 edition of Falloppia's lectures on tumors, by Donato Bertelli in Venice, did not touch upon a whole series of clinical pictures that can be found in the relevant section of the 1606 edition of Falloppia's *Opera omnia*, and Agathus' notes on Falloppia's lectures on the first book of Dioscorides' *Materia medica* were a mere fragment, dealing only with a handful of substances.[29]

If historians have shown very limited interest in the published lecture notes they have all but ignored the manuscript student notes on Falloppia's lectures that have come down to us. During my research in German, Austrian, Italian, and English archives and libraries, I discovered quite a number of such original lecture notes.[30] Most of them are not mentioned in Favaro's extensive biography or by later authors. The often cursory or indeed sloppy handwriting, with words crossed out that are then reinserted later in the sentence, brief additions, and the frequent rendering of Falloppia's words in the first person ("I have experienced", "I have found") and of his addressing his audience as "iuvenes" or "domini" leave no doubt that these are authentic testimonies of Falloppia's teaching activity.

These manuscript notes lend strong support, in turn, to the assumption that the published ones reliably document Falloppia's teaching, too. A comparison does not reveal any significant differences in structure, content, and character from the manuscript notes. Some of the published notes were particularly detailed but this can be explained by the fact that the publishers undoubtedly preferred to publish the most elaborate notes on a specific lecture they could get hold of.[31]

Falloppia was not only a professor and lecturer but also a renowned medical practitioner. Yet we so far know next to nothing about Falloppia's medical practice, about his preferred explanations and his diagnostic and therapeutic approach. Apart from the fact that medical historians have generally tended to neglect actual medical practice in favor of the major theoretical debates, hardly any sources were known to have survived in Falloppia's case, except for a couple of epistolary *consilia* Falloppia wrote for individual patients. However, much better and richer sources for Falloppia's diagnostic and therapeutic approach have so far been totally overlooked. Falloppia not only repeatedly mentioned cases from his own practice in his lectures but, more importantly, also frequently took part in the so-called *collegia*, for which Padua was famous. In a *collegium*, three, four, and sometimes even more professors discussed a concrete case in front of the students. First, the patient's medical history and current condition were presented. Then the professors, one by one, offered their diagnosis and their explanation of the causes of the illness in question and gave their therapeutic advice. Over thirty collegia in which Falloppia participated are documented in the posthumous 1587 edition of the *Consilia* of Vittore Trincavella, one of Falloppia's colleagues in Padua,[32] and I have found about two dozen others that are documented in manuscript notes in Siena and Wolfenbüttel.[33]

Many physicians in the sixteenth century maintained a lively epistolary correspondence with colleagues and often also with scholars who were active in other fields. Academic physicians were, in fact, major participants in the *res publica literaria* of their time. Given his fame and his numerous students, Falloppia undoubtedly also wrote and received numerous letters. Unfortunately, only very few of them have come down to us. Pericle di Pietro's 1970 edition of Falloppia's correspondence offers the transcripts of altogether thirty-five letters written by Falloppia, including a fair number of letters to Ulisse Aldrovandi in Bologna whose correspondence has been preserved in Bologna.[34] Two other letters, to Girolamo Giunti and Girolamo Mercuriale, which were published in the sixteenth and early seventeenth century already are not ordinary letters but epistolary treatises.[35] Only nine letters others wrote to Falloppia are known to have survived. With the exception of a letter from Mattioli,[36] which he published in his *Epistolae*, all of them are by officials or administrators.

In the course of my own research, I have found five more, hitherto unknown letters by Falloppia. Two of them he wrote in January and December 1550 from Pisa to the ducal *auditore* Francesco Torelli in Florence and to

the ducal secretary Giacopo Guidi in Livorno.[37] Also from his Pisa days, a
letter to Pietro Vettori (1499–1585), a renowned humanist and professor of
Latin and Greek in Florence, has survived. Falloppia requested Vettori's
help in finding out which shape the speaker's podium in the forum and the
senate in ancient Rome had and whether they were made of stone or wood
and whether there was a railing.[38] The Vatican Library in Rome holds the
copy of an epistolary consilium by Falloppia for an unnamed patient who
suffered from impotence.[39] Finally, Di Pietro overlooked a printed dedicatory epistle which Andreas Patricius addressed to Falloppia in 1558.[40] In
addition, I have been able to trace, in the Staatsbibliothek in Berlin, the
original of a letter Falloppia wrote to Giovanni Francisco Canani in Ferrara. Campori published this letter in 1864 but by the time Di Pietro assembled his edition, the original letter could no longer be found and he had to
rely on Campori's edition, which is rather faulty as it turns out.[41]

In addition to Falloppia's own correspondence, letters from medical students in Padua and other contemporary scholars are an important source
for Falloppia's biography. Here I have profited from the extensive database
on the correspondence of (German-speaking) physicians of the sixteenth
and seventeenth centuries, which we have built up in Würzburg since 2009.
Among the more than 55,000 letters listed in the database so far, there are
several dozen letters from German-speaking students reporting from Padua
in Falloppia's times and from Italian physicians and naturalists to German-
speaking scholars, writing about Italian and Paduan affairs. The most noteworthy collection in this respect are the detailed and entertainingly gossipy
letters from Georg Purkircher in the Trew-collection in Erlangen.[42]

Structure and organization

The book begins with a biographical sketch. The sources about the first
three decades of Falloppia's life up to his time in Padua are sparse and fragmentary. To this day, we do not know exactly where and how Falloppia acquired his medical knowledge and his anatomical skills. Even his teaching
activities in Ferrara and Pisa are poorly documented. In view of the countless contradictory and false claims about Falloppia's biography in historical
research, the main task here will be to point out what we actually do know
or for what there is at least some evidence. The sources are much richer for
his years in Padua, from 1551 onward, the time when he became famous for
his anatomical demonstrations but also as a surgeon. These most fruitful
and successful years of his professional career, which came to an abrupt end
with his early death in 1562, will stand at the center of this book.

My examination of Falloppia's teaching and research will begin with the
area in which he acquired the greatest fame: anatomy. I will first look at
Falloppia's anatomical lectures and at his public and private anatomical
demonstrations on humans and animals. Given that the utility of anatomical knowledge for the practice of a medicine that was based on notions of

humors, spirits, faculties, and *intemperies* is far from obvious, I will also highlight his consistent efforts to link anatomical knowledge and medical practice. Furthermore, I will discuss the claim historians continue to make to this day that Falloppia performed – or was at least accused of – practicing vivisection. While the historical evidence does not support this claim, the printed and handwritten student notes on Falloppia's anatomical teaching leave hardly any doubt that Falloppia, by his own admission, gave deadly doses of opium to criminals who had been sentenced to death and were sent to Pisa by the Tuscan authorities to be killed and dissected there.

Falloppia's anatomical research and discoveries have understandably attracted particular attention among historians and in the medical world. Falloppia himself duly emphasized most of these discoveries in his *Observationes anatomicae* of 1561. In some cases, student notes make it possible to date individual discoveries more precisely, well before 1561. In addition, some discoveries whose value Falloppia did not explicitly emphasize in the *Observationes* can be found in student notes. They offer the earliest known description of the ileocaecal valve, for example, an anatomical structure near the caecum that prevents a reflux of fecal matter from the colon into the small intestine. The discovery was later attributed to the Swiss anatomist Caspar Bauhin (1564–1624), who studied among others in Padua. Falloppia demonstrated its function to his students by filling the colon of a monkey with water.

The third chapter is devoted to Falloppia's surgery. A detailed examination of the numerous surgical topics Falloppia covered in his lectures is beyond the scope of this study and would be of limited interest to most readers. For the most part, Falloppia relied on the relevant ancient and contemporary surgical literature, on Guy de Chauliac's work in particular, supplementing it only, especially with regard to therapy, with his own contributions and experiences. I will limit myself therefore to an outline of the major topics and of Falloppia's significance as one of the most famous early representatives of surgery practiced by *doctores medicinae*. In addition, two topics on the fringes of surgery will be discussed in some detail because Falloppia had a pioneering role here: medical cosmetics and the *morbus gallicus* or "French disease". In the context of the French disease, Falloppia described among others a linen sheath, which had to be soaked in various medicines, dried, and put over the glans to prevent the French disease. This passage, which was first published in 1563, has widely been taken to represent the first description of a condom. Based on handwritten student notes, I can date this invention to the 1550s already but a closer reading also shows that this "condom" was much too small for use during intercourse and was to be applied – for hours in fact – afterward only, as a prophylactic treatment.

Falloppia's botanical and pharmaceutical interests stand at the center of the fourth chapter. Contemporaries already praised Falloppia as a second Dioscorides. He was friends with Luca Ghini, the founder of the botanical garden in Pisa, corresponded with Ulisse Aldrovandi in Bologna, later one of the most famous naturalists of his time, and shared his house for years with Melchior Wieland, who was eventually appointed prefect of the botanical garden in Padua. Today, various species of buckwheat of Asian origin are named after Falloppia, among them *Fallopia multiflora* and *Fallopia japonensis.*

My account of Falloppia's botanical activities will nevertheless be rather short. It would not make sense to present his detailed discussion of individual medicinal plants. By contrast, I will study in more detail Falloppia's lectures on metals, stones, and earths, and on thermal springs. Here Falloppia combined natural philosophical discussions, about the source of the heat and the mineral admixtures in the healing waters and on the generation of stones and metals, with an explanation of the therapeutic benefits. In this context, I will also describe the chemical analyses that Falloppia performed. Moreover, looking at his theoretical statements on the action of medicines and poisons through their "total substance" (rather than through their peculiar mix of elementary qualities), I will highlight Falloppia's role as a major representative of medical empiricism. By contrast, the often made claim that Falloppia conducted experiments on criminals who had been sentenced to death to test the effects of opium on the human body will be shown to result from a misinterpretation of the historical sources.

In the following chapter, I will examine Falloppia's practical work in medicine and surgery, mainly on the basis of the handwritten and printed student notes that document the opinions Falloppia expressed on the occasion of the so-called *collegia* in Padua, where various professors offered their judgment on a specific case in front of a student audience.

Having analyzed Falloppia's activities in and his contributions to these various fields, the book will return to Falloppia's biography. I will describe his uneasy position in the professorial hierarchy in Padua, which probably was the reason why he sought to leave Padua and assume a different professorship in Bologna. I will discuss, in some detail, his special relationship with Wieland, with whom he lived for several years and who was at the center of a heated public controversy with the famous botanist Pietro Andrea Mattioli, who then also attacked Falloppia. And I will tell the story of his last disease and death and that of the murder of Bassiano Landi, a major opponent of Falloppia in Padua, who was said to have rejoiced at Falloppia's death and who was fatally wounded, just a few days after Falloppia's funeral, possibly by followers of Falloppia.

The concluding chapter will look at Falloppia's legacy and, in particular, at the various editions of the published student notes on his lectures on different topics, at the editions of his *Opera omnia* by competing publishers in Venice and Frankfurt, and at the *Secreti*, which was by far the most widespread and popular of all the works that appeared under Falloppia's name but of which he was not the author.

I have listed the different editions of Falloppia's *Observationes anatomicae* and of the student notes on his lectures in a separate "Bibliography of Falloppia's Works", which is organized according to three major subfields:

anatomy, surgery, and pharmacology. Other printed primary sources and the research literature will be found in the "General bibliography". I have dispensed with the often arbitrary and inconsistent capitalization of individual words, especially in Latin book titles, and have modernized the punctuation. The translations of quotations in Latin and Italian are my own, unless explicitly stated otherwise. Since these quotations often contain subtle linguistic nuances, which the translation cannot fully capture, I frequently give the original phrasing in parentheses or in the notes.

Notes

1 In his study on the "anatomical Renaissance", Andrew Cunningham (Anatomical Renaissance (1997)) devoted only a few lines to Falloppia, in striking contrast to his extensive treatment of Andreas Vesalius, Realdo Colombo, and Girolamo Fabrizi d'Acquapendente. Andrea Carlino's *Books of the body* mentions him once only, as the teacher of Volcher Coiter (Carlino, Books (1999), p. 224). He also plays a marginal role only in Cynthia Klestinec's study of anatomical teaching in post-Vesalian Padua (Klestinec, Theaters of anatomy (2011)), although Falloppia and his student and successor Fabrizi d'Acquapendente were the most important and influential anatomists in Padua at the time.
2 Favaro, Gabrielle Falloppia (1928).
3 Montalenti, Gabrielle Falloppia (1923), p. 59; the publication carries the date 1923 but it must have come out much later: the note on p. 59, which was added when the proofs were being prepared, refers to Favaro's work of 1928 and to a review of that book by Montalenti in the same year.
4 Fantuzzi, Memorie (1774), with an edition of eight letters from Falloppia to Ulisse Aldrovandi (pp. 194–217); Campori, Lettere (1864); Corradi, Tre lettere (1883); Angelini, Una lettera (1900); Raimondi, Una lettera (1903); De Toni, Spigolature X (1913).
5 Di Pietro, Epistolario (1970).
6 Falloppia, Observationes (1964).
7 Di Pietro, Contributo (1974); Belloni Speciale, Falloppia (1994).
8 See the bibliography.
9 Mortazavi et alii, Fallopio (2013).
10 An illustrative example of this kind of writing which comes in the garb of a scholarly (and peer-reviewed) paper is Öncel, One of the great pioneers (2016). On just a couple of pages, the author claims falsely among others that Falloppia first studied the "classical sciences" before he "moved on to priesthood", that it is "uncertain" whether he studied there with Andreas Vesalius in Padua, that he "performed vivisections during his 3 years in Pisa", "dedicated" [sic!] his "Observationes Anatomice" [sic!] to Petrus Manna, described the function [sic!] of the ovarian duct, and "produced [sic!] simple condoms" made of linen and "tried them on 1100 men".
11 Calderato, Brevi cenni (1862), note on p. 8, already refuted this claim, quoting the relevant passage from Falloppia.
12 Such errors can even be found in the writings of some well-established medical historians. Pietro Capparoni (Capparoni, Profili (1928), pp. 46–49) was one of those who did not even get the year of Falloppia's death right and wrongly claimed that Falloppia performed vivisections on humans. Arturo Castiglione, in his times a leading expert in Italian medical history, wrongly claimed that

Falloppia – who fell sick and died eight days later – died "after a long illness" (Castiglione, Fallopius (1962), p. 183).

13 For example, the entry on "Fallopius" in the *Biographical dictionary of scientists* (Palmer, Fallopius (1994)) wrongly asserts that Falloppia was taught by Vesalius in Padua; the entry on "Falloppio" in the *Oxford dictionary of the Renaissance* claims that the Tuscan Duke Cosimo de' Medici invited him to Padua (which was not under Tuscan rule) in 1551 (Cosimo called him to Pisa in 1548), uncritically repeats the unsubstantiated claim that Falloppia was accused of vivisection, and misquotes Falloppia as having asserted that he examined the genitals of 10,000 patients with syphilis; similarly the *Dictionary of medical biography* (Ongaro, Falloppia (2007)) claims, without any evidence, that he studied medicine in Modena, graduated in Ferrara in 1547 (the correct date is 1552), and that he was accused, in Pisa, of practicing vivisection. The entry in the *Enzyklopädie Medizingeschichte* (Tshisuaka, Falloppia (2005), pp. 391–392) claims that he was a student of Vesalius in Ferrara (where Vesalius never taught), that he was appointed as a professor of anatomy in Ferrara in 1548 (he never was), and that he spent the year 1560 in Paris (he was there for a few months).

14 The Italian version wrongly claims that Falloppia studied with Colombo in Padua and was a student of Giovanni Battista da Monte (https://it.wikipedia.org/wiki/Gabriele_Falloppio, accessed 24 August 2021). The English version gives the year of his doctoral degree as 1548 (rather than 1552), wrongly claims that he was professor of anatomy in Ferrara and that he was appointed in Pisa in 1549 (rather than 1548), attributes the publication of the notes of his students exclusively to Volcher Coiter, and naively takes Falloppia's claim seriously that he tested the "condom" on 1,100 men in "an early example of a clinical trial" (https://en.wikipedia.org/wiki/Gabriele_Falloppio, accessed 24 August 2021). The German Wikipedia (https://de.wikipedia.org/wiki/Gabriele_Falloppio, accessed 22 December 2021) reproduces two of the errors about Falloppia's time in Ferrara in the *Enzyklopädie Medizingeschichte* (see note above) adding that he was the first to mention the French disease in a scientific treatise (dozens of treatises had been published before).

15 For an early summary of Falloppia's major anatomical findings, see Haller, Bibliotheca anatomica (1774), pp. 218–221; more recent examples are Wells, Fallopio (1948); Kothary and Kothary, Fallopio (1975).

16 The most noteworthy exceptions are Gurlt, Geschichte der Chirurgie (1898), pp. 361–403, with a detailed summary of Falloppia's surgical teaching; Casoli, Sifilografi (1905), pp. 19–28, on Falloppia's account of the French disease; Zanier, Medicina e filosofia (1983), pp. 12–19, on Falloppia's work on metals; Ferrari, L'opera idro-termale (1985) and Hsu, Gabrielle Falloppia's 'De medicatis aquis' (1993), on thermal springs; Gadebusch, Medizinische Ästhetik (2005), pp. 86–95, on Falloppia's *De decoratione*.

17 Tiraboschi, Biblioteca modenese (1782), pp. 236–253; Calderato, Brevi cenni (1862); Favaro, Gabrielle Falloppia (1928); Di Pietro, Epistolario (1970).

18 Observationes anathomicae. In: De ossibus (1570), foll. 71r–76r.

19 Stolberg, Teaching anatomy (2018).

20 Niedersächsische Staats- und Universitätsbibliothek Göttingen (henceforth SUBG), Ms. Meibom 20; the last detailed and datable notes in this manuscript are on the lecture *De partibus similaribus* which Falloppia began on 4 December 1552 and on Falloppia's dissection of a dog and a monkey in January and February of 1553; on fol. 248r, the writer entered a few lines on the dissection of a human uterus which he witnessed on 18 January 1554. He only briefly listed the structures he had seen, with no clear reference to the "tubae". Considering the great detail in which the writer recorded Falloppia's anatomies in 1551–1553, it

remains uncertain whether these notes refer to Falloppia's teaching; the writer may have moved on to study somewhere else.

21 The manuscript seems to have come into the possession of Johannes Sigfridus (1556–1623), who taught anatomy at the University of Helmstedt in the 1580s and 1590s. Sigfridus published a new edition of Falloppia's *Observationes* in 1588 (Falloppia, Observationes (1588)). Toward the end of the manuscript, a loose slip of paper – which may well have originally been placed elsewhere in the manuscript or indeed at the very beginning – is styled as a frontispiece, with the words, written in a rather hasty hand:

> Ex ore ipsius Falloppii olim inter dissecantem et demonstrantem exceptae et diligenter conscriptae. Nunc vero in gratiam studiosorum in 5 libros digestae et luce donatae opera et studio Johan. Sigfridi etc. Accessit Anatomia simiae ab eodem Falloppio administrata et habita.

In his edition of the *Observationes*, Sigfridus arranged the material in five books but there is no mention of the anatomy of a monkey; on the manuscript, see also Blumenbach, Nachricht (1783), pp. 372–374.

22 Edited in Mache, Anatomischer Unterricht (2019); on Fracanzano see Anastasio, Fracanzani (1997).

23 Österreichische Nationalbibliothek, Wien/Vienna (henceforth: ÖNBW), Cod. 11210. On Handsch and his medical and poetic activities, see Senfelder, Handsch (1901); Smolka/Vaculínová, Renesanční lékař (2010); Stolberg, Empricism (2013); Storchová, Handsch (2020); Stolberg, Learned physicians (2022), which also draws extensively on Handsch's notes.

24 Falloppia, Compendium (1571); cf. Chapter 2.

25 Puccinotti, Storia (1859), p. 642, described them harshly as "tutte impasticciate daisuoi discepoli e intarsiate di teoriche, che al Fallopio erano affatto o sconosciute o disapprovate".

26 On Agathus, who seems to have used the name Giovanni Bonacci as a pseudonym, see Mazzuchelli, Gli scrittori, vol. 1,1 (1773), pp. 177–178, vol. 2,3 (1762), p. 1530.

27 Dedicatory letter from Bruno Seidel to the rector and the professors of Marburg University, 1 March 1577, in Falloppia, De ulceribus (1577).

28 Ibid.: "Etenim sententias me integras et perspicuas expreßisse non dubito, et phrases autori familiares conseruasse, tametsi forte alicubi verbum unum atque alterum breuitatis et festinationis causa inter scribendum additum aut detractum reperietur, quod et solet et aliter non potest fieri."

29 Falloppia, Opuscula (1565), foll. 23v–33v.

30 See the list of manuscript sources at the end of this book.

31 An exception is, to some degree, *De partibus similaribus*, published by Falloppia's student Volcher Coiter in 1575 (Falloppia, De partibus (1575)). As Coiter explained in the introduction, the text was based on the notes that Georg Marius and Joachim Camerarius had made, in different years, on the lectures Falloppia had delivered on the topic. Coiter clearly reworked and edited the text. He not only inserted an additional chapter (chapter 6). Falloppia also does not appear as a speaking "I" but in the third person and in the context of the veins Vesalius' reply to the *Observationes anatomicae* is cited which was published after Falloppia's death.

32 Trincavella, Consilia (1587).

33 Biblioteca comunale, Siena, Misc. XVI, CIX 32, foll. 1r–15v, notes by an unidentified student on three *collegia* in which Falloppia expressed his opinion, in March and April of 1559; Herzog-August-Bibliothek, Wolfenbüttel, Cod. Guelf. 22 Aug. 4°, foll. 1v–130v, notes by an unidentified student on *collegia* he witnessed in Padua in 1555/56.

34 Di Pietro, Epistolario (1970); see also Biblioteca Universitaria, Bologna, Mss Aldrovandi 98¹, fol. 61r, "Herbae petitae a Falopia quae reperiuntur in horto patavino".
35 Falloppia, De asparagis (1565); Falloppia, De obstructionibus (1615).
36 Mattioli, Epistolarum (1561), pp. 159–170, 1 January [1558].
37 Archivio di Stato di Firenze, Fondo Guidi 571.
38 British Library, London, Add. 10266 (vol. 5, alphabetical order, no pagination), letter from Falloppia to Vettori, Pisa, 30 September 1550, in Italian.
39 Biblioteca Vaticana, Rome, Regin. lat. 1297, pp. 191–193.
40 Patricius, Dedicatory epistle (1558).
41 Staatsbibliothek Berlin, Sammlung Darmstaedter 3c 1550, Padua, 1 April 1561; cf. Campori, Lettere (1864).
42 See www.aerztebriefe.de.

1   Biography

Modena

Before embarking on a sketch of Falloppia's biography, I should briefly explain my spelling of his name, which may be unfamiliar or seem outright wrong to some readers. In historical writing and in library catalogues, his name is often written as "Gabriele Fallopio" or "Falloppio". By all appearances, this spelling derives from the Latin book titles and other contemporary sources that referred to him as "Gabriel Falloppius" or "Fallopius".[1] Like Favaro, Di Pietro, and others before me, I prefer the spelling which Falloppia himself used: he signed his (Italian) letters consistently with "Gabrielle Falloppia". "Fallop(p)ia" – and not "Falloppio" – was also the name by which his family was known in Modena. Agostino Gadaldin, a major figure in the intellectual life of Modena, likewise referred to him as "il Falloppia".[2]

Falloppia's exact date of birth is not known. Some writers have claimed that he was born in 1490 but this is clearly wrong.[3] Others have given 1523 as his year of birth and this date has been widely accepted in most recent historical writings and biographical encyclopedias.[4] The date is based on contemporary sources, which state unanimously that Falloppia died in early October of 1562, before reaching the age of forty.[5] At closer analysis, this makes 1523 only a possible year of Falloppia's birth, however. He could also have been born already in October, November, or December of 1522, as an entry by the *cancelliere* Gian Maria Barbieri in the minutes of the council in Modena may indicate. According to Barbieri, Falloppia died "not yet having completely arrived at the age of 40" ("non giugnendo ancora compitamente a i 40 anni di sua età").[6] In sum, all we can say is that Falloppia was very likely born in late 1522 or in early 1523.

Falloppia is usually assumed to have been born in Modena because this was where his family lived and where he seems to have grown up. Yet we cannot exclude the possibility that he was born out of town or even spent the first years of his life elsewhere. As a grown-up, he later told the story how he witnessed a farmer's son who was bitten by a snake in the countryside.[7] The baptismal registers preserved in the archives of the parish of *Cattedrale di*

DOI: 10.4324/9781003242000-2

Modena have an entry on the baptism of Gabrielle's younger brother Giulio Ludovico, on 4 October 1524, but none on Gabrielle.[8]

Drawing, in particular, on two contemporary chronicles of Modena, by Lodovico Castelvetro and Tommasino de' Lancilotti, Favaro has assembled the available information.[9] The father, Girolamo called Girvò, was a well-known figure in Modena. He led a soldier's life but also worked as a goldsmith for some time. His son later mentioned in a lecture that his father was present at the siege of Naples in 1494/95.[10] Eventually, he entered the service of Cardinal Ippolito d'Este and finally that of Count Guido Rangone. Surviving letters from his hand indicate that he had enjoyed at least a basic education. He became quite affluent and was able to buy some land. In 1521, he married Caterina de' Bergomozzi (d. 1557), whose family was well-established in Modena. Gabrielle, named after his grandfather, was their firstborn child. In 1527, Girvò is said to have supervised works on the construction of the fortress in Modena. In 1529, he was able to buy more land. In 1532, he was again on a military campaign but in the summer of 1533 he died in Venice, reportedly of dropsy ("intropixia").[11]

Gabrielle thus lost his father when he was about ten years old. Notwithstanding Caterina's dowry and the considerable possessions of land his father had accumulated, Girolamo left no fortune to Gabrielle and his younger brother Giulio (c. 1525–1550),[12] if the chronicler Castelvetro is to be believed.[13] In the historical literature, it has even been asserted that Falloppia grew up in poverty but there is no convincing evidence for this claim. One wonders, in fact, how his father could have lost not only his money but also his landed property and Falloppia himself later explicitly stated that poverty never beset him in such a way that he could have secured the comforts of life only with difficulty.[14] Even if Girolamo did not leave him much, the maternal relatives, especially Caterina's brother Lorenzo Bergomozzi, could support him and probably took care of Gabrielle in the following years. Lorenzo had gained an influential position in Rome, first as a singer and soon, it seems, as a personal confidant of Pope Leo X (1475–1521), and had been richly endowed by the Pope.[15]

At any rate, Falloppia must have enjoyed a good education. Modena, with a population of about 15,000–18,000[16] in the 1530s, was by contemporary standards an urban center with a lively intellectual environment. The famous humanist and historian Carlo Sigonio (1524–1584), his compatriot, later reported that he knew Falloppia from their early years when they were both taught by the same teachers.[17] One of these teachers was the already mentioned humanist scholar and chronicler Ludovico Castelvetro (born around 1505), who returned to his hometown Modena in 1529 after years of studies in Bologna, Padua, Ferrara, and Siena.[18] Both Falloppia and Sigonio learned Greek with Francesco da Porta (or Porto), a native of Crete, whom the city of Modena had hired as a Latin and Greek teacher and who later also taught at the University of Ferrara.[19] Da Porta's teaching activity is documented for the period from 1536 onward. Falloppia was about thirteen years old at the time, an age at which other later humanists and physicians attended a Latin school. He probably had already learned some Greek before. As Giovanni Grillenzoni, a leading figure in the intellectual life of Ferrara, later reported, Castelvetro and Giovanni Falloppia – presumably a relative – had found someone with some knowledge of Greek who offered daily lessons in Grillenzoni's house already before Da Porta's arrival.[20]

Undoubtedly, Falloppia also came into contact with the members of the so-called *Accademia modenese*, a circle of humanists that formed in those years around Grillenzoni. In the late 1530s, the *Accademia* and its members came under the suspicion of adhering to "heretical", that is, Lutheran or Calvinist beliefs. There was talk of a "Lutheran sect" in Modena. He had heard that many of them ate meat on Friday and neither prayed nor fasted on the Sabbath, a vicar complained to the bishop of Modena at the time. According to the vicar, Gabrielle Falloppia, in particular, was considered a heretic in Modena.[21] Falloppia's name, along with that of others, is also found in the Inquisition files, in a list with altogether fourteen names of suspected heretics that were reported from Modena to the *Sacrum Officium*.[22] As a result of these proceedings, those suspected of heresy in Modena had to sign the forty-one articles of a declaration of faith ("formulario di fede") in 1542. Among them were three cardinals and other clergy and laymen, as well as Lorenzo Bergomozzi and Gabrielle Falloppia. The latter confirmed with his own signature that he affirmed all the articles and that he submitted himself to the Holy Catholic Church in Rome.[23]

In their study of Falloppia's religious beliefs, Monica Panetto and Vito Terribile Wiel Marin concluded that his initial humanist interests came to converge with Protestant convictions and in the end, they claim, Falloppia adopted Calvinist beliefs. It is not clear, however, to what degree the suspicions of the Church about the humanists of the Modenese *Accademia* were justified – leave alone those regarding Falloppia who was not even twenty years old at the time. There certainly is no conclusive evidence. The vicar himself had to admit that he could not find anyone who could help him prove that Falloppia held heretic views.[24] Grillenzoni, the leading figure in this so-called *Accademia*, defended himself and the men around him against the accusations, arguing that they were just a circle of scholars, without any formal statutes. They only cultivated the study of Greek and Latin and, he insisted, had never read a single word from the Bible in their meetings. He attributed the "calumnies" against them to the Dominican friars who resented the emergence of another center of learning and who took it badly that Grillenzoni had protested against the burning of a simple old woman as a witch.[25]

If the evidence from the inquisition is inconclusive, some of the other evidence Panetto and Wiel Marin cite for Falloppia's Calvinist leanings is outright unconvincing. They cite a letter from Falloppia to Ulisse Aldrovandi in which he expressed his concern which had gone lost.[26] As we will see, he was in negotiations with Bologna at the time, however, because he wanted to leave the University of Padua. Padua was under Venetian rule and the Venetian authorities were not to know about his plans. Panetto and Wiel Marin also interpret Falloppia's "empirical-rational" approach as pointing to Reformed convictions. But this turn away from a focus on the study of authoritative works toward the study of nature was a general phenomenon in sixteenth-century medicine and natural history and it was supported and promoted also by Catholics and Lutherans.

What is certain is that Falloppia appeared and was perceived for years as a clergyman in Modena. In 1545, Tomaso Lancilotti in his *Cronica di Modena* reported that Falloppia walked around in the garb of a clergyman.[27] In February 1547, he was explicitly referred to as a "clericus et mansonarius" of the church in Modena, when his uncle Lorenzo Bergomozz gave him and a certain Laurentius de Amatoriis his house with a pharmacy ("apotheca") in Modena in gratitude for their (unspecified) services. A *mansonarius* was a kind of sexton, someone who took care of a house and more specifically of a church but we do not know, what duties this position entailed in Falloppia's case.[28] A month later Lorenzo also gave up his canonry in favor of Falloppia, which Falloppia renounced, in turn, a few months later.[29] Finally, in October 1548, Lorenzo bequeathed him a parish in Villa Montale. When Lorenzo died in April 1549, his canonry fell again to Falloppia by inheritance. He ceded it to his uncle Francesco Falloppia in August 1549.[30]

Young Falloppia thus probably made his living in the service of the Church or from ecclesiastical benefices but by his early twenties he must have turned to medicine. By December of 1544, Lancilotti already described Falloppia as studying medicine more than acting as a clergyman.[31] There is unequivocal evidence that he performed a public anatomical demonstration at the Ospedale della Morte in Modena in late 1544. At the beginning of December, the *conservatori* approved a request of the *collegium medicum* for financial support for "unam anatomiam" to be undertaken for the instruction of the young doctors of the city. A few days later, the *collegium* was granted 10 libra for an anatomy of the body of Giovanni Battista Cimino from Agro Piceno, who was to be executed the following day, 13 December 1544.[32] Young Falloppia dissected the body in public. Lancilotti, who was otherwise quite critical of him, praised Falloppia's demonstration as excellent.[33] And so did Castelvetro: "Without a teacher, he devoted himself to the study of the herbs and to cutting human bodies", he claimed, and without ever having seen the dissection of a human body, "he dared to dissect one publicly in Modena, and he showed better than others what is usually shown".[34]

Around the same time, Falloppia is said to have started practicing medicine. In his *Cronica di Modena*, Lancilotti mentions three patients by name whose injuries Falloppia treated in August and October of 1545 and who all died. Lancilotti added that Falloppia had never studied medicine, that he had only practiced medicine since Christmas (i.e. around the time of the public anatomy) and had never practiced surgery.[35] Some historians have taken Lancilotti's stories to prove that Falloppia's lack of medical knowledge and experience caused the death of the three men.[36] All three apparently had suffered serious injuries, however, and Lancilotti was by no means a neutral observer. Twenty-eight-year-old Achille Carandino, who died in October 1547 under the treatment of Falloppia and that of various barber-surgeons, was Lancilotti's adoptive son ("legitimato"). And he not just doubted Falloppia's skills but also those of all the physicians and surgeons in Modena except for a certain "maestro Augusto da Cavola", who did know how to cure injuries like the one Achille had suffered.[37]

It seems extremely unlikely, in retrospect, that Falloppia gained his medical knowledge as a mere autodidact, as Castelvetro and many authors after him have claimed. He would hardly have been called and entrusted with the treatment of serious injuries without any previous study and training and it seems virtually impossible that Falloppia would have been able to acquire the knowledge and skills he needed for conducting a successful public anatomy without having witnessed others dissect a human body before. The neat images of the human body and its parts in illustrated anatomical textbooks are deceptive. Anyone who has dissected a corpse knows how difficult it is to identify and separate the individual anatomical structures, the layers of tissue, and the vessels and nerves that run through them. It is a messy business.

This raises the question where Falloppia could have acquired the necessary medical and anatomical training. There was no university in Modena at that time where he could have studied.[38] Almost certainly he had access to learned medical writings. In the *Observationes anatomicae*, which he addressed to the physician Pietro Manna of Cremona, Falloppia mentioned the Modenese physician Agostino Gadaldini as a "most learned" man, who had done great service to their "common studies", clearly referring to medicine.[39] Gadaldini (1515–1575) was the son of a bookseller and printer in Modena and a leading exponent of medical humanism. In 1541/42, he published the Guintine edition of Galen's collected works, which offered largely up-to-date, humanistic translations of Galen's works. Gadaldini had entrusted the revision of the translations of Galen's anatomical works in this edition, and in particular that of Guinther of Andernach's translation of Galen's *Anatomical procedures*, to Andreas Vesalius.[40] Gadaldini had an excellent library, including various Greek manuscripts.[41] For a young man like Falloppia who had enjoyed a good education in the classical languages, it would seem obvious that he should seek contact with a man like Gadaldini. A letter Gadaldini wrote to Ludovico Castelvetro in 1553, reporting that he and Falloppia had so far not been able to obtain works by Petrus Ramus on rhetorics, indicates that Gadaldini and Falloppia remained in contact also during Falloppia's Padua years.[42]

In one of his lectures, Falloppia later also mentioned an uncle on his father's side ("patruus") by the name of "D. Petrus Agathus", who was a skillful "investigator of natural and supernatural things". Falloppia still had some of his works and manuscripts in his possession and quoted his instructions for the preparation of *talcum*.[43]

Reading learned medical treatises was hardly a sufficient preparation to undertake the treatment of patients, however, leave alone a public anatomical demonstration. There were a number of learned physicians in Modena, however, who may have acted as Falloppia's medical teachers or mentors and whom he may have accompanied on their house calls. This was not uncommon at the time. Young Georg Handsch, for example, followed Ulrich Lehner in his practice in Prague, when he was about twenty years old, before he went to Padua to study medicine.[44] Even the students in Padua, Bologna, Montpellier, and Ferrara, the leading centers of medical education at that time, owed their practical knowledge and skills to a considerable degree to the opportunities these places offered them to accompany their professors and other experienced physicians on their visits to patients, in the hospital and in private homes.[45]

A major medical figure in Modena was Niccolò Macchelli, who played a leading role in the local *collegium medicum*. Macchelli was a learned, humanist physician. He produced, among others, a new Latin translation of the Greek edition of Rhazes' treatise on the plague and a treatise on the *morbus* gallicus, which he addressed to the young members of the Modenese *collegium medicum*.[46] The circle of scholars in the *Accademia* in Modena also counted the physicians Giovanni Villanova and Alessandro Baranzone among its members and its *spiritus rector*, Giovanni Grillenzoni, was a trained physician as well.[47] Obviously, there may also have been other learned physicians in Modena – not to mention barber-surgeons and other practitioners – with whom Falloppia may have been in contact and from whom he may have learned.

Falloppia later told his students repeatedly about cases in Modena. A particularly elaborate story of his was about a young nobleman in Modena who had an artery of the head severed by an injury. The man bled very profusely and lost a lot of blood. Doctors and surgeons were called to him from Modena and other places and tried everything possible to stop the bleeding. Falloppia was present, too. When even cauterizing the artery twice did not have the desired effect, he suggested that the artery be tied off. This was not the usual practice at the time and the "other physicians" ("medici") – implicitly, Falloppia already counted himself among the physicians here – disagreed. According to his own account, Falloppia did not have the courage to stand up against them because he was still young and inexperienced. When the doctors and surgeons retired into another room to discuss how to proceed and whether to cauterize a third time, Falloppia went back to the patient, however. As he palpated the wound, he accidentally placed his finger on the pulsating, severed artery. He immediately had a string given to him and tied off the artery, stopping the bleeding – to the amazement of the physicians and surgeons.[48]

As to anatomy, it has often been claimed that Falloppia studied with Vesalius, who taught in Padua and also performed anatomical demonstrations in Bologna and Pisa in the early 1540s. However, there is no evidence whatsoever for this claim. It is based on a misreading of a passage in the *Observationes anatomicae*, where Falloppia stated that he was one of the "school" of Vesalius. He added immediately that this was because he read his writings carefully, the same way in which Vesalius was a student of Galen, not because he heard him speak *viva voce* but because he studied his works.[49] Falloppia must have found other opportunities to acquire the skills he needed to perform a successful dissection in 1544 and that earned him the trust of Macchelli and the *Collegium medicum*. He may have had a chance to witness other physicians perform autopsies on deceased patients and he may have practiced on animals but, as we will see in the following chapter, the most likely explanation, at this point, would seem that he undertook medical and anatomical studies in Ferrara several years earlier than historians have so far believed.

Ferrara (1540/45–1548)

Ferrara, the capital of the Duchy of Este, was not far from Falloppia's hometown Modena. In the Convent of San Domenico, one of the most famous universities in Europe at that time attracted students from far and wide.[50] With Niccolò Leoniceno, Giovanni Manardi, and Antonio Musa Brasavola, leading representatives of medical humanism were active in Ferrara in the early sixteenth century.[51] Moreover, along with Padua, Ferrara was in those years a leading center of anatomical studies.[52] Both developments were closely related. The intensive scrutiny of the ancient medical writings by the medical humanists gave crucial new impulses to empirical anatomical research. On the one hand, the texts that were rediscovered offered numerous findings that had gone lost; on the other hand, the work of Galen, the major ancient anatomical authority, contained numerous errors, which could only be assessed and if necessary corrected by autopsy. Vesalius' epochal *De humani corporis fabrica libri VII* offers the best and most famous example.

With Giovanni Battista Canani (1515–1579), one of the leading practicing anatomists of the time was active in Ferrara in the 1540s.[53] He conducted public anatomies and, in addition, held anatomical demonstrations in his own house.[54] Around 1541, before Vesalius' epochal work appeared, Canani (also known as Canano) published the first book of his exquisitely illustrated *Musculorum humani corporis picturata dissectio.*[55]

From 1543 to 1552, Canani also taught surgery and practical medicine in Ferrara.[56] Unlike at the universities north of the Alps, where it was practiced almost exclusively by artisans as a craft, surgery already had a firm place at the Italian universities by the early sixteenth century.[57] Canani seems to have been a very skillful and experienced surgeon who also devised surgical instruments of his own.[58] The scope of his surgical practice went far beyond the treatment of simple wounds, ulcers, and rashes, which was the daily bread of barber-surgeons. Amatus Lusitanus, who taught in Ferrara from 1542 onward, reported, for example, on the case of a two-year-old child with a – presumably congenital – obstruction of the urethra. The urine flowed out through an opening at the base of the scrotum. To cure the child, Canani, according to Amatus, had a fine silver tube ("cannula") made, with a silver needle inside. The cannula was to be inserted into the orifice in the scrotum and, apparently following the natural path of the urethra, pushed along the penis. When the cannula encountered a blockage, the needle would perforate the impeding structure and permit the urine to flow through its natural exit.[59]

Returning to Falloppia, it is uncontroversial that he studied in Ferrara at some point in the 1540s. In his lectures, he later repeatedly mentioned and praised Antonio Musa Brasavola as his teacher. Unlike at universities north of the Alps, the study of medicine in Italy did not usually require a previous degree as *baccalaureus* or even *magister* in the *artes liberales*. Medicine was taught within the arts faculties. For a young man like Falloppia, who thanks to Francesco da Porta's teaching (and presumably also to that of Agostino Gadaldini) had a solid humanist education and a very good mastery of Greek, Ferrara must have seemed an obvious choice. It was the capital of the Duchy to which Modena belonged and Antonio Musa Brasavola was one of the most famous medical humanists and botanists of the time.[60] In geographical terms, the nearby University of Bologna would also have been an option but there is no evidence that Falloppia studied there. He always named only Brasavola in Ferrara as his teacher, and the entry for his later doctorate in Ferrara – he was already teaching in Padua at the time, after a few years in Pisa – explicitly mentions his times at the universities in Ferrara, Pisa, and Padua but not in Bologna.[61]

The crucial question remains when exactly Falloppia started studying in Ferrara. Unfortunately, the old archive of the university with the documents from the time before 1620 has not survived.[62] Based on circumstantial evidence, Favaro and other historians with him have assumed that Falloppia studied in Ferrara, at the earliest, in the academic year 1545–1546, when he was already twenty two or twenty three.[63] The reason why they have rejected the possibility that Falloppia may have studied medicine (with Brasavola) and surgery and anatomy (with Canani) in Ferrara before that time seems to be that until then Falloppia occasionally figured in sources from Modena and also held an ecclesiastical office there. Ferrara is only about forty miles from Modena, however, and we do not know to what degree Falloppia's office entailed concrete duties. Moreover, most of Falloppia's specific, datable activities in Modena mentioned in the sources refer to the summer and fall, before the university lectures started.[64] It is moreover striking that Falloppia later mentioned the case of a severely wounded "gipsy" ("zingarus") whom he had seen as an "adolescens" in Ferrara.[65] The term "adolescens" usually referred to the age prior to that of the "iuvenis", the term Falloppia, following prevailing usage, routinely used when he addressed his own students.

In sum: the evidence is not conclusive but given the considerable anatomical skills he must already have acquired by the end of 1544, it would seem not only possible but quite likely that Falloppia began to study medicine in Ferrara earlier than it has so far been believed.

This would also help explain why the university had appointed him as a lecturer by the academic year 1547/48. Falloppia, who had no doctoral title yet, taught on medicinal plants ("ad lecturam simplicium medicamentorum"), probably reading and commenting on *De materia medica* by the Greek naturalist Dioscorides.[66] The date 1547/48 comes from the earliest datable evidence of Falloppia's teaching: the *rotuli* (salary lists) and a mandate of 31 July 1548 specified the salaries the various lecturers had earned in the academic year that had just ended.[67] Since no contract or official letter of appointment has survived, it is possible, however, that Falloppia started teaching earlier already. There are some hints in this direction. In the *Observationes anatomicae*, Falloppia reported that Canani had discovered a small, hitherto unknown muscle in the palm of the hand, today's *musculus palmaris brevis*, "when I was teaching in Ferrara (it was almost 13 years ago)".[68] Since the *Observationes* appeared in 1561, this could indicate again 1547/48 but Falloppia stated in the introduction that he had written the book four years earlier, and Pietro Manna, whom Falloppia addressed directly in the *Observationes*, had already died in 1560. A posthumous eulogy by Melchior Wieland, who lived with Falloppia for years and probably knew him better than anyone else, points into the same direction. According to Wieland, Falloppia had taught for eighteen years at the universities of Italy. Since Falloppia died in the fall of 1562, this would imply that he began his teaching career in 1544.[69] If Wieland's figure is correct, Falloppia would initially almost certainly have taught without an official lectureship, however: from October 1546, another mandate with the payments due to the various lecturers has survived, which refers to a "D. M. Gasparem Gabrielis" from Padua "ad lecturam simplicium medicamentorum" – presumably Falloppia's predecessor as a lecturer on the simples.[70] Given the anatomical skills which Falloppia had already acquired by the end of 1544, when he performed a public anatomy in Modena, he may have taught anatomy at first or at least assisted Canano in his anatomical teaching before obtaining Gaspare Gabriele's lectureship on the simples.

Issues of chronology apart, it is certain that Falloppia studied medicine in Ferrara and it was here, years before he had even received his doctorate, that he started his successful career as a university lecturer. He learned anatomy from the leading anatomists of his time and conducted a public anatomy and several private dissections himself in Ferrara.[71] Musa Brasavola, his teacher in medicine, was a renowned practitioner and a leading expert on medicinal plants. And Falloppia's occasional references to patients he saw others diagnose and treat in Ferrara suggest that he also received some bed-side training in practical medicine.[72]

Pisa (1548–1551)

Falloppia soon must have made a name for himself as an anatomist, not only in Modena and Ferrara. In the fall of 1548, Falloppia accepted the invitation of the Duke of Tuscany, Cosimo de' Medici, to teach anatomy at the university in Pisa and began teaching a few weeks later.[73] Unfortunately, we know very little about Falloppia's time in Pisa. Favaro and Di Pietro have undertaken extensive searches in the local archives in Pisa and in Florence but found no evidence of Falloppia's activities. Lorenz Gryll, who stayed in Pisa for nine months around 1550, mentions Falloppia in his travel report only as "rei anatomicae peritissimus" without going into more detail.[74]

All we know is that Falloppia developed extensive anatomical activities in Pisa. According to the notes of his student Georg Handsch, Falloppia later told his students in Padua of nine people who had been sentenced to death and who were made available to him so he could dissect them.[75] Two letters I have recently discovered in the archives in Florence indicate that Falloppia had indeed been assured that the government of Cosimo I would secure a sufficient supply of corpses. They also suggest, however, that things did not always go smoothly. On 5 January 1550, Falloppia wrote to Francesco Torelli, the ducal *auditore* in Florence seeking his support. Sant'Antonio (17 January) was near and it was time to begin with the anatomy. There was already one "subject" at hand but one corpse did not last very long. He asked Torelli to give orders that another "subject", female or male, be made available to him. If that proved impossible, he begged Torelli to make a customs official, the *proveditore di dogana*, buy a couple of monkeys and a pregnant goat.[76] In November of 1550, at the beginning of the following academic year, he turned with a similar request to the ducal secretary Giacopo Guidi. Christmas time was approaching, when he usually performed his anatomical demonstrations, and he did not know whether there would be corpses ("soggetti") available. He needed at least one "healthy body", he explained, otherwise no good anatomy would be possible. He had heard that a few days ago some persons had been brought from Barga – a town north of Lucca – to Florence to be executed and asked for one of them. Last time, he complained, he had been served poorly. They gave him an old man who had been sick with quartan fever for months and whose body decayed ("corruppe") in a flash. It was also essential to have a monkey but last time it was only when the anatomy was over that he received the letter telling him about the monkey that had been assigned to him. He asked that he be given that monkey earlier this year and that, if that was not possible, another monkey be procured.[77] We do not know how successful Falloppia was with his request this time.

Falloppia's activities as a medical and surgical practitioner in Pisa are also documented only very vaguely, mostly by occasional references and anecdotes we find in student notes on his lectures. To illustrate the power of negative emotions, for example, Falloppia told the story of a student from Lucca who suffered a very slight cut with a knife but was in such great fear that Falloppia had a very hard time curing him.[78] In another lecture, he used the example, from his Pisa years, of a boy who had fallen with his head on a sharp stone to explain the surgical treatment of superficial head injuries. The case gave rise to a controversy among the physicians and surgeons whether the wound in the scalp which was in part no longer attached to the skull should be sutured or not. He knew from many other cases, Falloppia assured his students, that it was better in such cases to suture the wound.[79]

Falloppia sometimes must have traveled also to other parts of the Duchy during his time in Pisa. In his Padua-lectures, he repeatedly mentioned various spas around Pisa, especially the ones near Lucca, which he had seen.[80] From Pisa, Falloppia also went on botanical excursions. Discussing the use of salsaparilla against the French disease, he told his students of a specimen of *smilax aspera* which he had found on the slopes of Monte San Giuliano (today's Monte Pisano, a few miles from Pisa). He had someone dig the plant out for him and successfully used it on patients with the French disease.[81] In his lectures on the simples, he also mentioned plants that could be found in the hills and mountains around Pisa, *asphalatum* for example.[82]

Apparently, he also spent some time in Florence and at the court of Cosimo I. In his lecture *De partibus similaribus*, he claimed that he had examined a hundred lion bones in Florence and found them to contain marrow.[83] And discussing the treatment of the French disease with salsaparilla, he told his students about a Spanish merchant who brought Cosimo four salsaparilla plants, complete with the leaves, fruits, and roots; Falloppia found that he had been wrong so far and that they were similar to the *smilax aspera* Dioscorides had described.[84] In another lecture, he described a plant that the Florentine Giovanni Battista Pedaldo had brought with him from the East Indies.[85]

Sometimes he also treated patients in Florence. Once he cured a boy from the nobility there whose cheek and eyelid had been badly cut and bruised by the hoof of a horse. According to Hippocrates, Falloppia explained, he should not have sutured the wound in the eyelid. He nevertheless did suture it because otherwise the eye would have been deprived of its cover. This would let the cornea harden and ultimately lead to blindness. His treatment was successful.[86] In his letter to the ducal *auditore* Torelli in Florence, he also gave Torelli some dietetic advice and recommended medicine for his ailing eyes, which would "preserve the sight and prevent the flux of subtle matter".[87]

Padua (1551–1562)

Falloppia remained only for three years in Pisa. On 9 September 1551, he was appointed as a lecturer at the University of Padua.[88] He still had not obtained a doctoral degree; it was only on 3 October 1552 that he received his medical doctorate under Antonio Musa Brasavola in Ferrara.[89] After Vesalius' departure from Padua, Realdo Colombo (1516–1559) had first taught anatomy and surgery there, followed in 1547 by Giovanni Paolo Giudizio from Urbino.[90] In his notes on an anatomical demonstration by Antonio Fracanzano, which he witnessed in the winter of 1550–1551, Handsch also mentioned Alessandro Veronese as the one who did the actual cutting but Veronese may have been a practicing surgeon and not a lecturer.[91]

According to the lists of Paduan lecturers and their salaries, Falloppia had to lecture on surgery and simples and the task of dissecting corpses. His salary was initially 200 fiorini per year and was raised to 270 fiorini in 1560 only.[92] He was to work and teach in Padua until his early death in 1562, spending most of his time there. We only know of two occasions, for sure, on which he was absent for longer periods of time. In the spring of 1552, he went to Rome, with the approval of the Venetian authorities, to treat sick Baldovino del Monte, the elder brother of Pope Julius III, together with Giambattista Canani, who may have recommended him.[93] By September 1552, he was back in Padua and his appointment was renewed.[94] With the beginning of the new academic year 1552/53, he began to teach again. A second longer absence is documented for the spring of 1560, when Falloppia ended his lectures early.[95] He obtained the permission to accompany and secure the medical care of a Venetian embassy to the royal court in Paris.[96]

In Padua, Falloppia reached the high point of his professional career. He made significant anatomical discoveries, attracted numerous students, and ran a successful practice. In the following chapters, I will look at his activities in these various fields in detail, before I return to Falloppia's biography, his final years, the controversies he got involved in, his death, and his legacy.

Notes

1 Canalis, Gabrielle Falloppia (2018), p. 175.
2 Staatsbibliothek Berlin, Slg. Darmstaedter 3d 1553, letter, Venice, 27 October 1553. In modern Italian, the word "faloppa" is still in use. It can refer to the cocoon of a silk worm that decomposes or putrefies. The coat of arms used by Falloppia's ancestors shows three little balls which historians have taken to represent such cocoons (Castiglioni, Fallopius (1962), note on p. 182). "Faloppa" also means "loudmouth" or "braggart", however, which would seem the more probable origin of a man's name.
3 For example, Mercklin, Lindenius renovatus (1686), p. 311.
4 Du Chastel, Vitae (1618), p. 206; Fabrucci, De pisano gymnasio (1760), p. 108.
5 According to Zamoscius, Oratio (1562), fol. 7v, he died "annos XL. natus", that is, in his 40th year; according to Wieland (Wieland, Papyrus, p. 111), who lived with Falloppia for years and must have known him very well, he died not yet forty years of age ("non annos natum quadraginta").
6 Di Pietro, Contributo (1974), p. 63; Di Pietro interpreted the passage correctly as indicating late 1522 or early 1523 but nevertheless considered the claim "acceptable" that Falloppia was born in 1523 (ibid., p. 65).
7 Falloppia, De tumoribus (1606), fol. 21v.

8 Di Pietro, Contributo (1974), p. 66; the baptismal registers of the other parishes in Modena are not extant. In theory, the family could also have moved from another parish in Modena into the cathedral parish between the birth of Gabrielle and that of Giulio.

9 Favaro, Gabrielle Falloppia (1928), pp. 26–35 and pp. 203–213 (excerpts from Lancilotto's handwritten *Cronaca di Modena*).

10 See also Falloppia, De morbo gallico (1563), fol. 2r, "cum fieret illa maxima obsidio Vrbis Neapol[is] ubi pater meus affuit."

11 Favaro, Gabrielle Falloppia (1928), p. 35.

12 Ibid., pp. 212–213.

13 Ibid., pp. 34–35.

14 Falloppia, Observationes (1561), ad lectorem: "neque rerum egestas ita me oppressit, ut tam difficili, ancipitique ratione vitae commoda fuerint mihi paranda."

15 On Lorenzo Bergomozzi see Favaro, Gabrielle Falloppia (1928), pp. 37–42.

16 Beloch, Ricerche (1908), p. 5.

17 Sigonio, Emendationum (1557), introductory letter to Francesco Robortello, 5 September 1557: "me quem ipse iam inde ab ineunte pueritia probe noverat, cum ambo ab iisdem magistris erudiremur, atque in iisdem studiis ita versaremur, ut alter alterius industria incitaretur." Sigonio taught Greek in Modena, from 1546 until 1552, then went to Venice and Padua where he renewed his controversy with Robertello.

18 Marchetti/Patrizi, Ludovico Castelvetro (1979).

19 Vedriani, Dottori (1665), p. 169; Borsetti, Historia (1735), vol. 2, p. 166.

20 Casoli, Statuti (1911), p. 73.

21 Firpo/Marcatto, Il processo (1984), pp. 936–937, "il qual intendo che si confetta in queste heresie"; see also Panetto/Wiel Marin, Falloppia (2001), pp. 276–277.

22 Archivio di Stato di Modena, Elenco di libri e di delati, busta I, ed. in Panetto/Wiel Marin, Falloppia (2001), p. 304.

23 Panetto/Wiel Marin, Falloppia (2001), esp. pp. 276–291; Favaro, Gabrielle Falloppia (1928), p. 52.

24 Quoted in Firpo/Marcatto, Il processo (1984), pp. 936–937: "Ma io non ho persona che mi dia indicio fermo per reprenderlo."

25 Cit. in Casoli, Statuti (1911), pp. 73–74.

26 Panetto/Wiel Marin, Falloppia (2001), p. 290; they are referring to the letter from Falloppia to Aldrovandi, 30 January 1559, ed. in Di Pietro, Epistolario (1970), pp. 47–48.

27 Cit. in Favaro, Gabrielle Falloppia (1928), p. 210.

28 Archivio notarile di Modena, Memoriali notarili dell'anno 1547, L. 2, N. 20, fol. XIIIr-v, cit. in Favaro, Gabrielle Falloppia (1928), pp. 214–215.

29 Favaro, Gabrielle Falloppia (1928), p. 53.

30 Ibid., pp. 53–54; Favaro also published some of the notarial documents.

31 Cit. ibid., pp. 209: "che studia in l'arte medicina, più che d'esser prete."

32 Archivio storico comunale di Modena, Atti della comunità, 1544 (no pagination), cit. in Favaro, Gabrielle Falloppia (1928), pp. 213–214.

33 Cit. in Favaro, Gabrielle Falloppia (1928), p. 200: "fece un'anatomia per eccellenza."

34 Cit. in Tiraboschi, Biblioteca (1782), p. 239.

35 Cit. in Favaro, Gabrielle Falloppia (1928), p. 210: "Il suo medico è stato D. Gabriel Falloppia, che non ha mai praticato ne studiato in medicina, e va vestito da prete, il che è da Natale in quà, che cominciò a medicare"; ibid., "che non ha mai praticato cirurgia".

36 Ibid., p. 61.

37 See the passage from Lancilotti's *Cronica* on 15 October 1545, quoted ibid., pp. 210–211. In the case of another of the three patients, Lancilotti himself saw the fault with other physicians who continued to treat the patient as well.
38 Foucard, Documenti (1885), p. 40.
39 Falloppia, Observationes (1561), fol. 76r.
40 Cf. Fortuna, Latin editions (2012); Galen, Opera omnia (1541), vol. 1, 58v–103v; the text is preceded by Latin translations of two other texts, which Vesalius had seen through, namely, on foll. 49v–50v, *De nervorum dissectione* and, on foll. 50v-55r, *De venarum arteriarumque dissectione*.
41 Petit, Gadaldini's library (2007); Nutton, Medical humanism (2021), n. 13.
42 Staatsbibliothek Berlin, Slg. Darmstaedter 3d 1553, Venice, 27 October 1553.
43 Falloppia, De morbo gallico (1563), fol. 25r.
44 Stolberg, Learned physicians (2022).
45 Stolberg, Bed-side teaching (2014).
46 Machelli, De morbo gallico (1555); idem, Razae libellus (1555).
47 Casoli, Statuti (1911), p. 64.
48 Falloppia, De vulneribus (1606), pp. 377–378.
49 Falloppia, Observationes (1561), fol. 4r.
50 Survey in Münster, Die Universität zu Ferrara (1968).
51 Castellani, De vita (1767); Petrucci, Vite (1833), pp. 79–85; Nutton, Rise (1997).
52 See Muratori/Menini, Contributi (1946).
53 Barotti, Memorie (1793), pp. 138–147; Petrucci, Vite (1833), pp. 91–96; Zaffarini, Scoperte (1909); Glabbeck/Biesbrouck, Giovanni Battista Canani (2020); on Canani's teaching in Ferrara see Muratori, Academic career (1969).
54 Muratori, Academic career (1969), pp. 313–316.
55 Canani, Musculorum (ca 1541).
56 Muratori, Academic career (1969), pp. 313–316.
57 Nutton, Humanist surgery (1985); McVaugh, When universities (2016).
58 Zaffarini, Scoperte (1909), pp. 22–24.
59 Amatus Lusitanus, Curationum (1551), pp. 158–159. In the end, the parents, fearing for the child's life, did not allow Canani to perform the operation.
60 Nutton, Rise of humanism (1997); Gliozzi, Brasavola (1972).
61 Pardi, Titoli dottorali (1901), p. 164.
62 Muratori, Academic career (1969), p. 311.
63 Tiraboschi Biblioteca modenese (1782), p. 239; Münster, Universität (1968), p. 59.
64 In this sense, Favaro argued that Falloppia began to study medicine in Ferrara in the academic year 1545/46 only, because he treated a patient in Modena in October 1545 (Favaro, Gabrielle Falloppia (1928), p. 67).
65 Falloppia, De vulneribus (1606), p. 254.
66 Dioscorides, De medica materia (1547).
67 Borsetti, Historia (1735), vol. 2, pp. 170–171; Muratori/Menini, Contributi (1946), p. 65 and p. 60 (photographic reproduction of the mandate of 1548 in the Archivio comunale di Ferrara); Di Pietro (Epistolario (1970), p. 21) also found a request, signed by Falloppia, that the local *giudice* was to pay 8 lire from Falloppia's salary to the *bidello* (Biblioteca comunale, Ferrara, Autografi, Racc. Cittadella, n. 1084). Di Pietro assumes that it was written in the academic year 1547/48 but the request carries no date.
68 Falloppia, Observationes (1561), fol. 102v.
69 Wieland, Papyrus (1572), p. 110: "Salve itidem, qui feliciter et exercuisti medicinam, et luculenter docuisti in clarissimis Italiae gymnasijs annos XIIX."
70 Muratori/Menini, Contributi (1946), p. 60; Gabriele also figures in previous mandates from 1538 and 1543 (ibid., pp. 57–58).
71 Münster, Universtität (1968), p. 59.

72 For example, Falloppia, De tumoribus (1563), fol. 38r, on a female patient with podagra in Ferrara; Falloppia, De bubone pestilenti (1566), fol. 14r, on various Portuguese patients in Ferrara, in whom he witnessed the effects of bezoar.
73 Letter from Falloppia to the secretary of Cosimo I de' Medici, Ferrara, 6 September 1548, Archivio di Stato di Firenze, Archivio Mediceo del Principato 390, fol. 61r, ed. in Favaro, Gabrielle Falloppia (1928), p. 215.
74 Gryll, Oratio (1566), fol. 5v.
75 ÖNBW, Cod. 11240, fol. 78r.
76 Archivio di Stato di Firenze fondo Guidi 571, letter, in Italian, from Falloppia to Francesco Torelli, Pisa, 5 January 1550.
77 Archivio di Stato Firenze, fondo Guidi 571, letter from Falloppia to Giacopo Guidi, Pisa, 6 November 1550:

> Questi giorni scrissi al S[igno]r Lolio [?] ricordandoli che s'auicina la Pasqua di natale nella quale facciamo l'Anatomia, ne sapiamo anchora se ui sono soggietti o non. Mi rispose che cio toccaua a V[ostra] S[ignoria] alla quale scriuerebbe et che io douetti negotiar' seco poi che era cosi vicina presente. Ho aspettato il ritorno di V[ostra] S[ignoria] et poi che non uiene et gia siamo sotto le feste ho pigliato questo spediente di racordarli con questa mia il bisogno nostro, perche se non habbiamo almeno un corpo che sia sano, non si fara cosa buona. Odo che di Barga furono condotti certi a Firenze alcuni giorni sono per giustitiarsi. Quella uedra se se potesse hauerne uno. L'anno passato stemmo molto male, che ci diedero uno uecchio che haueua hauuta la 4na molti mesi, et si corruppe in un tratto. Appresso non possiamo far' cosa buona senza simia per le cose di Galeno, hora l'hanno passato ci fu assegnata una che ha ex[?] Lucca Martini. Ma non ci furono date le l[ette]re prima che fu fornita l'anatomia. Pero V[ostra] S[ignoria] sara contenta di commetter' che ci sia data a buon' hora quest'anno, et non ci potendo hauer' questa ne faccia procacciar' un'altra. Accio che possiamo far' compiutamente et con diligenza il n[ost]ro ufficio, per far' honor (come merita) al n[ost]ro Ill[ustrissi]mo padrone. Non altro. N[ost]ro S[igno]re Dio la feliciti. In Pisa il VI Nouembre 1550.

78 Falloppia, De vulneribus (1606), p. 389.
79 Ibid., p. 319.
80 Falloppia, De medicatis aquis (1564).
81 Falloppia, De morbo gallico (1563), fol. 39v.
82 Falloppia, De materia medicinali (1606), p. 243.
83 Falloppia, De partibus similaribus (1606), p. 105; "a hundred" stands undoubtedly for "many"; such figures must not be taken by the letter, in the writings of Falloppia as in those of other contemporary writers.
84 Falloppia, De morbo gallico (1563), fol. 39v
85 Falloppia, De materia medicinali (1606), p. 243.
86 Falloppia, De vulneribus (1606), pp. 360–361.
87 Archivio di Stato di Firenze, fondo Guidi 571, letter from Falloppia, Pisa, 5 January 1550.
88 Tomasini, Gymnasium patavinum (1654), p. 96; Favaro, Gabrielle Falloppia (1928), p. 216.
89 Pardi, Titoli dottorali (1901), p. 164.
90 ASV, Riformatori allo Studio di Padova 449, fascicle *Lettori dello Studio di Padova dal 1542 sino al 1556* (copy made in 1614); Montalenti (Falloppia ([1928]) gives his name as Paolo Guiduccio.
91 ÖNBW, Cod. 11210, fol. 191r.
92 See the list in Favaro, Gabrielle Falloppia (1928), pp. 219–220, with the entries on Falloppia in the *rotuli* of 1551–1555, 1557 and 1562 (the *rotuli* of the other years

are missing) found in the Archivio Antico dell'Università di Padova, filza n. 651, foll. 193v, 199v, 207v, 211v, 226v and 241r and in the *bollettari* of 1551–1555, 1558 and 1560–1562, ibid., foll. 192v, 198r, 206v, 210v, 216v, 230v, 234v, 238v and 240v.

93 Favaro, Gabrielle Falloppia (1928), pp. 106–108.

94 Ibid., p. 216; ASV, Riformatori allo Studio di Padova 449, fascicle on *Lettori dello Studio di Padova dal 1542 sino al 1556* (copy made in 1614).

95 Falloppia, De vulneribus capitis (1566), fol. 59v; according to Biblioteca comunale Urbania, Ms 95, fol. 66r, he continued this lecture, in June 1561, with the parts on injuries that he had omitted for legitimate reasons the year before.

96 Decree of the Venetian Senate, 13 March 1560, ASV, Senato I. R o 42 Terra 1559–1560, agosto, foll. 123v–124r, ed. in Favaro, Gabrielle Falloppia (1928), p. 218; according to a letter from Falloppia's student T. de Rochefort to Jacques Dalechamps in Paris, which Falloppia was asked to bring with him, the Venetian ambassador was Bernardo Naugerio (Bibliothèque nationale, Paris, Ms. Lat. 13063, fol. 214r–v, Padua, 2 April 1560).

2 Anatomy

Renaissance anatomy has for a long time attracted the particular interest of historians. Many publications have described the major advances, praising Andreas Vesalius, in particular, and his beautifully illustrated *De humani corporis fabrica libri septem*. In recent years, various authors have significantly expanded the scope. Highlighting the influential and innovative work of Jacopo Berengario da Carpi (ca. 1460–1523) and other pre-Vesalian anatomists and the anatomical skills and findings of post-Vesalian anatomists, they have arrived at a more nuanced assessment of Vesalius' achievements who was far from alone, at the time, in criticizing certain aspects of Galen's work.[1] New questions have been asked as well. Historians are now looking more closely at the methods and the philosophical and theological background,[2] the teaching of anatomy,[3] and at the dramatic, ritual elements[4] and other aspects of the intellectual, cultural, and social history of anatomy.

Falloppia's fame in the history of medicine and science rests on his lasting contributions to the study of human anatomy, which he published in his *Observationes anatomicae*. In his own days, before the publication of the *Observationes*, Falloppia was renowned above all as a teacher, however. At the time, students from all over Europe flocked to Padua and other Northern Italian cities to enjoy a training in anatomy that was far superior to what the universities north of the Alps could offer. They were keen, in particular, on anatomical demonstrations, which allowed them to see, with their own eyes, the individual parts and the complex fabric of the human boy.

Lectures

Like Vesalius before him and in contrast to the traditional practice, to which post-Vesalian anatomists in Padua had returned in the meantime,[5] Falloppia combined the tasks of *lector, demonstrator,* and *sector* in one person in his anatomical teaching. This does not mean that he fulfilled the three roles at the same time. As the records of the Helmstedt Anonymus make clear, Falloppia first outlined in detail the *historia* of the structures he would show afterward on the corpse and explained their function.[6] Presumably, these lectures did not take place in the often freezing cold dissection room

DOI: 10.4324/9781003242000-3

or outside in the open air but in a building, quite possibly the university building. In these lectures, Falloppia also resorted to graphic representations. Thus, Falloppia showed the anatomy of the different layers of the eye ball first on the basis of an illustration ("in schemate").[7]

The student notes on these anatomical lectures resemble to a large degree those on other lectures. Falloppia quoted various authorities, first and foremost Galen and Herophilos and much more rarely and mostly critically Vesalius. He discussed the opinions of different authors, for example about the function of individual parts of the body, and expressed his own convictions, describing his own findings at the dissection table. These lectures must have taken a considerable amount of time. The notes on Falloppia's teaching, which the Helmstedt Anonymus made during Falloppia's first public anatomy in Padua, in January 1552, extend over nearly 200 pages.[8]

In addition to the anatomical lectures that preceded his anatomical demonstrations, Falloppia also gave lectures on individual anatomical texts and topics, without the use of a corpse. Compared to his numerous lectures on surgical and pharmaceutical topics, these lectures played only a minor role in his teaching, however. They mainly focused on topics that could easily be covered without dissecting a cadaver. He lectured, in particular, on *De partibus similaribus*, that is, roughly, on what we would call different types of tissues today – bones, fat, nerves, muscles, etc.,[9] – and on Galen's *De ossibus*.[10]

The earliest known record of Falloppia's lectures *De partibus similaribus* is found in the notes of the Helmstedt Anonymus.[11] It was largely a commentary on the relevant passages in Galen's work. Falloppia taught repeatedly on the topic. Bruno Seidel's edition of Falloppia's *De partibus similaribus* (1575) was based, according to him, on the notes that Georg Marius (presumably around 1556/57) and Joachim Camerarius, who studied in Padua from 1559 to 1561, had written on different occasions.[12]

Theoretical lectures on the bones do not seem to have attracted much interest among the students: "This lecture was heard by few" commented Helmstedt Anonymus on his notes on what Falloppia said about the bones of the pelvis and the lower extremities, at the end of his lectures on Galen's *De ossibus*.[13] The anatomy of the bones could be demonstrated conveniently by using dry and clean bones or indeed an entire skeleton. Skeletons were not easy to get hold of. Making a complete skeleton, first cleaning and boiling the bones and then putting them back together with wire, was a laborious and time-consuming process.[14] After his public anatomy in Modena, Falloppia had made such a skeleton. As Lancilotti reported in his *Cronica*, he had the bones boiled and then joined them together with glue and copper wire.[15] In Padua, there was already at least one such skeleton in 1550, before Falloppia's arrival; perhaps it was a legacy of Vesalius, who is said to have made several skeletons there.[16] In his notes on the first public anatomy he attended in Padua in the winter of 1550/51, Georg Handsch specifically mentioned a complete skeleton next to the head of the dissected corpse.[17] The skeleton was supported by a long, rounded "iron" in the spine that passed through the canal which normally contained the spinal cord. Later, Handsch also made notes on various bones he had seen in a skeleton ("Quae in skeleto vidi").[18] Apparently, Falloppia also demonstrated and explained animal bones to him and the other students. Handsch noted: "he also showed the skull of a dog and a monkey."[19]

Anatomical demonstrations

Anatomy brought a major innovation into the world of academic teaching. In the other disciplines, the written and spoken word was paramount and, as we just have seen, the teaching of anatomy also relied to a considerable degree on lectures. The principal reason, however, why anatomical teaching in the Northern Italian universities (and in Montpellier and Paris) was taken to be far superior to that elsewhere was that these universities also offered the students exceptionally good opportunities for seeing the anatomy of the human body and its parts with their own eyes. "Autopsia", seeing oneself, was not only a major new epistemological ideal. It was also at the heart of anatomical teaching.

Historians of Renaissance anatomy have mostly focused on the big public anatomical demonstrations, which increasingly took place in theaters that were built for that purpose. These public anatomies were major events in the life of the university and, as the *Acta* of the *Natio germanica* in Padua show, the students complained bitterly when they did not take place. A public anatomy might last for ten days or even longer and offered a unique chance to gain a systematic overview of the anatomy of the human body. Already a brief look at the size and design of sixteenth-century anatomical theaters makes clear, however, that these big public events were far from ideal when it came to teaching and showing the intricate structure of the body and its organs. Most spectators – and particularly the students who were confined to the upper ranks – would hardly be able to see the finer parts, for example the different vessels or nerves that entered or exited from an organ like the liver or the stomach.

For teaching the subtle anatomy of individual organs or parts of the body, the so-called *anatomiae privatae* were much more suitable. Some historians have misinterpreted these "private" anatomies as dissections the anatomists performed on their own, "in private", for the purpose of research. In contemporary usage, the term referred to anatomical demonstrations for students, however, which, in contrast to the public anatomies, were for a small group of (presumably paying) students only. Here the students could stand around the dissection table and approach the corpse, looking from a short distance at what the anatomist showed them. Sometimes they might even get a chance to do some of the manual work of dissecting themselves and thus develop skills that would be crucial if they were later asked – or volunteered – to perform a public anatomy or an autopsy on a deceased patient. Vesalius had already emphasized that it was important for students to try their hand at the task.[20]

Falloppia strongly recommended that his students follow Galen's advice and practice on animals first, before dissecting humans, so they would develop the necessary skills. The anatomy of the human body was particularly obscure and things would be difficult to find without practicing first on animals; just like good musicians first played a *praeludium* to make sure that their strings were in harmony.[21]

Falloppia's student Handsch not only wrote down some of the well-known basic practical rules, for example that the abdominal organs should be dissected first because they decomposed quickly.[22] He also made note of the various instruments the anatomist used for different purposes: various kinds of scalpels, a small knife – Falloppia's preferred tool – a probe that allowed him to explore branching vessels, sponges to absorb fluids, oil that made it easier to separate the muscles from one another, and the burning candles which generally served as a source of light at the time.[23] Handsch also found noteworthy how Falloppia started the dissection, first removing the hair (or fur in the case of animals) then making two cuts in the shape of a cross to open the abdomen, one down from the sternum to the pubic bone and one across the navel toward both sides, followed by folding back of the abdominal wall, initially only on one side.[24] Last but not least, he took note of some important tricks. For example, when one dissected a liver, it was good to first tie the veins with thread in order to prevent the blood from flowing out.[25]

The sources are rather silent on such details but presumably such private anatomies could only exceptionally or indeed never be performed on men (or women) who had been condemned to capital punishment. They were difficult to get hold of. In Padua as in Montpellier, the anatomists were also sometimes allowed to dissect the bodies of deceased patients. In Montpellier, deceased patients, rather than criminals who had been sentenced to death, even seem to have made up the majority of those whose bodies were dissected.[26] This has important implications for our understanding of Renaissance anatomy. The common notion – supported by images of muscular men like those in Vesalius' *Fabrica* – is that Renaissance anatomy relied largely on the dissection of healthy men who were still young or at least in mid-adulthood and had died from the executioner's hand. Deceased hospital patients were a much more heterogeneous group and women constituted a larger group here. Some of them might even be visibly pregnant and would not have been executed in this state if they had committed a crime but some died from natural causes. Hospital inmates were also of different ages, from the very young to the very old. Last but not least, depending on the disease, the dissection might show pathological changes in specific organs which could not normally be found in convicted criminals.

There is no systematic record of Falloppia's public and private anatomical demonstrations. We have to rely primarily on what we learn from student notes, from the scarce remainders of Falloppia's correspondence, and from the *Acta* of the *Natio germanica*, which are unfortunately far from exhaustive in this respect. The following summary of his anatomical demonstrations gives some impression but it is almost certainly not quite complete.

In the academic year 1551/52, Falloppia dissected (at least) two bodies. The anatomy lasted for more than two weeks, from 7 January until 23 January 1552, on most days in two sessions, one in the morning and one after lunch.[27] On 24 January, he complemented his demonstration with that of the female genitals and the fetal vessels of a pregnant sheep.[28]

The following academic year, Falloppia must already have started in late December of 1552. He was already busy with a public anatomy, when, on 29 December, someone brought him the head of a dolphin ("phoca") from the fish market in Venice to dissect.[29] The Helmstedt Anonymus also took notes on the dissection of a dog, from 31 January until 1 February 1553, and from 2 February until at least 10 February of a monkey[30] whom Falloppia had tied by the hands and feet and then drowned in water. Falloppia must have brought the monkey with him from Pisa: he explained the abundant fat under the skin by the fact that he had fed the monkey well for four years in his house and the monkey had moved little.[31] Presumably, these were private anatomies for a smaller circle of students. Falloppia showed the various anatomical structures, just as he did on human cadavers. The students were also able to see some of the latest discoveries here, the ileocaecal valve, for example, and the stapes in the ear, and Falloppia, as with his demonstrations on human cadavers, also addressed surgical questions.[32]

In the academic year 1553/54, Michinus witnessed the dissection of at least two bodies, in private and public anatomies.[33] The Helmstedt Anonymus only briefly reports the dissection of a uterus, on 18 January 1554, and it is not entirely clear whether he saw it in Padua.[34]

For 1554/55, only one public anatomy is documented and it ended quickly, after just a couple of days. The university – we will come back to that – had been instructed to follow the statutes and to let Falloppia only do the manual work of dissection and to leave it to Vittore Trincavella, the professor of practical medicine, to do the actual lecturing and demonstrating but the students protested so vehemently and made so much noise that Trincavella was not able to continue.[35]

In the following academic year 1555/56, Georg Keller, who studied medicine in Padua at the time, complained that there was no anatomical demonstration,[36] and in a letter to Ulisse Aldrovandi, Falloppia remarked, in May of 1556, that he had spent the whole winter, until carnival, sick in bed.[37]

In 1556/57, Falloppia combined his request to the *Riformatori allo Studio* in Venice for a human cadaver with the promise to make a most beautiful ("bellissima") anatomy on bears and on a monkey ("con gli orsi e la simia").[38] He seems to have performed at least one anatomy, for, in January 1557, he complained about a cold he got from it.[39]

For the following academic year 1557/58, we only have the much later, post-humous biography of Joachim Curaeus (1532–1573) by his friend Johannes Ferinarius (1534–1602) to go by. According to Ferinarius, who presumably relied on what Curaeus had told him decades before, an anatomical demonstration took place during the Christmas vacations. Ferinarius did not specify whether Falloppia also had the role of the lecturer on this occasion. He added that Falloppia was very diligent that year, dissecting seven bodies of humans and all kinds of animals.[40] If this is correct, Curaeus may well have witnessed private anatomies, which Falloppia performed with an eye on the *Observationes anatomicae* he was writing at the time.

In 1558/59, Falloppia mentioned an anatomy which had kept him from giving a letter to Aldrovandi to the messenger right away last time he came.[41]

In 1559/60, the representatives of the *Natio germanica* tried in vain. They saw the blame partly with Falloppia, who excused himself with his poor health, and partly with the rector, who for his part did not exert any pressure on him.[42] Understandably, the students suspected that Falloppia used his poor health as a mere pretext. Falloppia was in his prime, in his mid-thirties. He lectured extensively on a range of non-anatomical topics and also visited numerous patients. In April 1560, he ended his lectures early to travel to France with Venetian ambassadors.

In 1560/61, the *Natio germanica*, headed by Joachim Camerarius, took the initiative early on, to avoid that Falloppia again shirked this task "in his usual way" ("suo more"). This time they were successful and they praised his demonstration as quite careful and precise. Unfortunately, the weather conditions were poor and in the middle of the demonstration Falloppia cited his weak health again and brought the demonstration to a hasty end.[43] It was to be Falloppia's last public anatomical demonstration. In the following year 1561/62, the students repeatedly asked Falloppia for an anatomical demonstration. But he excused himself once more with his poor health and, on top of that, they complained, he had convinced the *Riformatori* that it was sufficient to perform a public anatomy every second year only.[44]

Anatomy and medical practice

The outstanding importance medical students in the sixteenth century attributed to anatomy in general and to anatomical demonstrations in particular, confronts the historian with a mystery. After all, the overwhelming majority of the future physicians who flocked to Padua and other northern Italian universities would eventually make their living as practicing physicians. Yet it is far from clear to what degree anatomical knowledge could contribute to a better understanding – leave alone a more precise diagnosis and more efficient treatment – of diseases. Medicine was dominated by humoral pathology which attributed most diseases to some morbid, foul, raw, slimy, or else burnt, acrid fluid matter in the body. Physicians could identify this matter far more easily in the excrements, in the stools and the urine above all, or in the bloodletting bowl than in the dead body. Sometimes this morbid matter might be ascribed to an *intemperies* of a specific organ, the stomach and the liver above all, but the diagnosis of a hot liver or a cold stomach did not require precise anatomical knowledge either.[45]

This retrospective assessment does not do justice, however, to the ways in which contemporary physicians and medical students perceived anatomy. In their eyes, anatomy was not just a quest for knowledge per se. Anatomical knowledge held great promises for a better understanding, diagnosis, and treatment of the many different kinds of ailments that befell the human body. Falloppia encouraged and fostered this belief. At one point, he even expressly declared his plan to discuss the diseases of the abdomen "so the use of anatomy and why it has to be dealt with diligently may be more obvious".[46] The knowledge of the nature, structure, and composition of the individual parts, their size, position, and their connections to neighboring parts was not only important for the purpose of knowledge as such ("scientiam ipsam"), he proclaimed in his surgical lectures. It was also indispensable for the art of medicine. In order to recognize pathological, preternatural changes in the body, one had to know the natural state "secundum naturam".[47]

In the case of injuries and surgical operations, the usefulness of detailed anatomical knowledge was quite obvious. Whoever wanted to understand and diagnose the injuries of the nerves and their possible consequences, Falloppia pointed out for example, had to be experienced in anatomy and know where nerves ran and where not.[48] Falloppia also stressed its importance for the understanding and treatment of internal diseases, however. The peculiar anatomy of the sixth cerebral nerve, for example, which reached down to the pleura, explained the intense pain of pleurisy.[49] He poured his scorn over physicians who claimed that it made a difference whether blood was let from the branches of the *vena humeraria* or those of the *vena axillaris*. Anatomy showed that both vessels were connected and thus both were equally suitable.[50] Showing the size and width of the gall bladder and the biliary duct, Falloppia explained that they were not easily obstructed. Jaundice could be caused there only by a stone or large amounts of sticky matter. By contrast, the tiny ducts in the liver itself could much more easily become obstructed.[51] Falloppia's discovery of the ileocaecal valve, which prevented a possible reflux of the feces from the colon into the ileum, had considerable implications, in turn, for one of the most commonly used therapeutic procedures. Galen had been wrong: the liquid of an enema and the various medicines that were added to it could not reach the small intestines leave alone the stomach.[52]

In his anatomical teaching, Falloppia sometimes also explained the pathological anatomy of diseases or mentioned findings he had made in postmortems of deceased patients. In the cadavers of patients with stones, his students learned, for example, he had found the stones in the renal cavity only not in the substance of the kidney itself.[53] Postmortems had also taught him that the pus which patients with pulmonary empyema excreted came from the lungs rather than from the pleura which surrounded them. He had found the pleura perfectly intact in empyematic patients. Rather than making a cut between the sixth and seventh rib to evacuate the pus, as the surgeons sometimes did, it was therefore better to promote the expectoration of the pus with suitable medicines. Cutting was advisable only in cases in which a swelling could be perceived from the outside.[54] Falloppia reported another autopsy in which he found what could be called a dropsy of the lung, namely a considerable amount of water that had collected on one side of the thoracic cavity. The pleura on that side was thickened and hard, he found.[55] Falloppia also told the students about his findings on the body of a girl with hydrocephalus: the water had collected between the brain and the *pia mater*, the soft lining of the brain.[56]

The number and range of instances, which the students documented, in which the anatomist underlined the uses of anatomy for a better understanding, diagnosis or treatment of internal diseases was not huge. It is clear, however, that Falloppia sought to make his students appreciate the importance of anatomical knowledge also for internal medical practice, apart from its value for surgery and natural philosophy. At the same time, with his references to the clinical, practical application of anatomical knowledge, he helped promote a major shift of focus in contemporary disease theory toward increasingly consistent attempts to identify the anatomical location, the area of the body or indeed the organ, where the pathological process had its principal site. We find a similar and closely related trend in bedside teaching – where Padua was at the vanguard – and in ordinary medical practice, where the manual examination of the patient's body and, in particular, of the abdomen acquired considerable importance since the 1530s.[57]

The anatomist as an executioner?

One of the great challenges of anatomical teaching in the Renaissance was to secure a sufficient "supply" of corpses. Even though they might sometimes be allowed or indeed asked to dissect the bodies of diseased patients, obtaining a suitable corpse exactly at the time when it was needed, namely in the winter vacations, when it was cold and no other lectures were taking place, could prove very difficult. And one corpse was not sufficient for an extended, careful anatomical demonstration, as Falloppia explained in his letter to Francesco Torelli. The organs and the flesh decomposed and putrefied too quickly.[58]

In Pisa, as we have seen, Falloppia could count on the support of the Duke and his officials. The authorities had orders to bring criminals who had been sentenced to capital punishment elsewhere to Pisa if necessary so that they could be killed and dissected there. Falloppia later told his students of (at least) nine people that he could dissect in his Pisa years.[59] Yet, as we have seen, things did not always go smoothly. In the winter of

1549–1550, he had to do with only one corpse for his anatomical demonstration, which on top that putrefied quickly because, as Falloppia believed, it was that of a man who had suffered from quartan fever from months.[60]

In Padua, Falloppia had the support of the Venetian government. In the spring of 1550, the Venetian authorities had threatened to inflict severe punishment on students who, against the divine and human laws, dug up corpses from their graves to perform anatomies and to sell – presumably for medicinal purposes – the fat and the bones.[61] The *Riformatori allo Studio di Padova*, who were in charge of the university, sought to secure other sources. But obtaining cadavers proved a challenge in Padua, too. In December 1556, for example, as the Christmas vacations approached, Falloppia himself approached the *Riformatori* for assistance. In his words, the good weather and the cold – there was snow – invited anatomy. The students urgently desired to see the fabric of the human body after having had to do without an anatomical demonstration for two years.[62] According to Falloppia, the German and Polish students were starting to lose faith in his promises and were already making moves to go to Bologna or Ferrara. He asked the *Riformatori* to make the *Podestà* in Padua either let him have as quickly as possible a "subject" or to allow the *massari*, the students who had been elected to assist in the anatomy, to secretly obtain the body of someone of low social status who had no acquaintances in Padua.[63] The *Riformatori* instructed the *Podestà* accordingly: if he had no "subject" at hand, who was about to be executed, he should allow the *massari* to secretly obtain the body of a "persona ignobile et non cognosciuta".[64] Apparently, they were successful.[65] In December 1557, the *Riformatori* wrote again to the authorities and rectors in Padua, requesting that the anatomists be supplied with "some subject", if the one they had gotten from Venice was not suitable, which they would find out from Falloppia.[66]

In the context of the supply of corpses for his anatomical demonstrations, Falloppia came under harsh attack from later writers. They accused him of a highly unethical, immoral behavior. According to some writers, Falloppia had defiled his name, in Pisa, by using criminals who had been sentenced to death, for experiments on the effects of deadly poison.[67] Others made an even more serious accusation. Falloppia, they claimed, performed vivisections on these criminals.[68] Scores of later writers have repeated these claims. Even Conde Parrado in a fairly recent and detailed analysis of ethical issues in Renaissance anatomy described Falloppia as using a criminal as a "guinea pig" ("cobaya") and accused him of a "macabre behaviour".[69] Others have further elaborated on such claims declaring – without any supporting evidence whatsoever – that Falloppia faced these accusations already while he was alive. O'Malley, for example, claimed that "Falloppia had the unpleasant experience of being accused, although wrongly, of practicing human vivisection".[70]

The accusations are based on two passages in the posthumously published student notes on Falloppia's lectures. Lecturing on preternatural swellings, Falloppia, according to the notes of an unknown student,[71] explained to his students that fever was opposed to the effect of a cold poison, which was the reason why one could fight such poisons, among other things by making the patient drink very hot wine. He related the following incident: "the prince" (meaning Cosimo I)

> orders that they give us a man, whom we kill and dissect in our own way. I gave him two drachms of opium. And the oncoming [fever] paroxysm (for he was suffering from a quartan fever) prevented the opium from taking effect. Boastfully he asked us to give it to him a second time, and that if he did not die, we would plead for his salvation with the prince. We gave him again two drachms of opium, outside the paroxysm, and he died.[72]

A second, similar passage is found in student notes on Falloppia's *De compositione medicamentorum*, also written by an unknown scribe and eventually published in print.[73] Here Falloppia discussed to what extent the primary qualities of medicines were effective *in actu* and, in this context, cited the action of opium – which was considered to be cold in a very high degree – on natural heat as evidence of such an effect. The notes relate the same story as above, only in other words and with some variations as to the quantities:

> when I was in Pisa, our Duke left the bodies to be delivered to justice to the anatomists, that they might give them the death which seemed to them [the most suitable]; we administered a drachm of opium to one, and killed him within seven hours; but we once administered [the opium] to another, who was suffering from quartan fever, whom the [feverish] rigor immediately seized, after the potion, which was then followed by the greatest heat, and he did not die, because the opium was overcome by the natural heat.[74]

This is all the evidence on which historians have based their accusations. There clearly is not even a hint at vivisection, and the story of the man with quartan fever does not describe an experiment either. Falloppia expected the first dose to kill the man who was then given a second dose when the first one did not have the desired effect.

The origin of these accusations is probably a passage in the work of the French physician Jean Astruc. In his historical survey of the history of the French disease, Astruc quoted the passage from *De tumoribus* correctly – and was outraged. Herophilus and Erastistratos had been guilty of even worse atrocities, dissecting criminals from prisons alive. Nevertheless, Falloppia's "barbaric cruelty" stunned him every time he just thought of it: a Christian doctor, in the sixteenth century, who did not shy away from openly taking the part of an executioner.[75] In other words, Astruc only compared what Falloppia had done with the vivisections the Alexandrian physicians had allegedly performed, and there is not a word of an experiment.

In the nineteenth century, Alfonso Andreozzi undertook extensive archival research on the supply of corpses in sixteenth-century Tuscany. He claimed that he had found evidence, beyond the passages quoted above, that Falloppia had performed vivisections on those condemned to death, with the consent of Cosimo I. The relevant chapter of his book is even expressly entitled "La vivisezione anatomica dei condannati a morte sotto Cosimo I duca di Toscana". A closer look at Andreozzi's sources shows that there is not even a hint at vivisection either. He probably misunderstood Astruc and read into his sources what cannot be found there. Andreozzi's findings are of great value for our understanding of the historical context, however. They confirm that the Duke ordered his local officials to supply the necessary cadavers for anatomical instruction in accordance with the statutes of the University of Pisa. If no cadavers were available in Pisa, the rector of the university was to turn to the Florentine judicial magistrate, the *Otto di guardia e balia*. The latter could send anyone who was sentenced to death in the whole of Tuscany to Pisa. In the Florentine archives, Andreozzi found documents on a series of instances between 1545 and 1570 in which the support the *Otto di guardia e balia* was indeed called upon for these purposes.

These documents also make it clear that those destined for anatomy were indeed sent alive to Pisa and handed over to the anatomist. In his work on Vesalius, Roth sought to exculpate the Pisa anatomists claiming that they only proposed the way in which those destined for anatomy would be executed.[76] But there is no mention of an ordinary execution in Pisa. The sources expressly state, for example, about a woman sentenced to death for infanticide on 14 November 1553, that she was brought from Firenzuola to the *Commissario* in Pisa at the end of December, at the time that is when the anatomical demonstrations usually began, and the *Commissario* in turn, as Andreozzi quotes from his sources, "handed her over to the anatomist as usual to dissect her".[77]

Some authors, in turn, have acknowledged the passages from the student notes, which state that Falloppia gave deadly amounts of opium to those he was about to anatomize but doubted that Falloppia had truly made this admission. They did not consider the students trustworthy.[78] There is no plausible reason, however, why the students should have invented this admission and the fact that it appears in the printed notes on two very different lectures weakens this argument. Additional and in fact the most important evidence against this argument comes from the surviving handwritten notes which document Falloppia's teaching in Padua, soon after his arrival from Pisa. They offer further, earlier instances of Falloppia telling his students about giving deadly doses of opium and these notes were for the students' personal use only. These students had even less reason to invent something Falloppia had never said. Georg Handsch, who studied with Falloppia in Padua from 1551 to 1553, repeatedly recorded Falloppia's remarks on this matter. Handsch's notes on Falloppia's lectures *De tumoribus praeter naturam* of 1552/53 do not specify that it was in fact Falloppia personally who gave the opium:

> For I saw a man condemned to death, to whom two drachms of opium were given in the attack of a quartan fever, and the fever overcame the poison and he was freed from it; on the second day opium was given to him again and he died as did the others.[79]

Handsch's most detailed entry on this story, however, states very clearly that Falloppia was not a mere spectator and that it was not a singular occurrence either. Indeed, it indicates that Falloppia gave the deadly poison even a lot more often than historians have assumed

> Falloppia told us of nine condemned men to whom he gave opium, to each two drachms in Malvasia wine with some *diacodium* (poppy syrup) to sweeten it. They took it and shortly after slept for four hours. Then they wanted something to drink and it was given to them, and they slept again, for eight hours, and breathed out [their souls] asleep.[80]

The story must have stuck in Handsch's mind. As a young physician in Prague, Handsch recalled again in his notebook this "hystoriam Fallopii", "who gave opium to nine condemned to death and one of them, with quartan fever, did not die".[81] In the context of a discussion with Pietro Andrea Mattioli, probably in November 1555, on how to make opiates more palatable to the patient, he noted again that Falloppia gave the opium to those sentenced to death together with *diacodium*, to sweeten it.[82]

The cases described by Andreozzi took place, with one possible exception,[83] before or after the period in which Falloppia taught in Pisa. However, there is hardly any doubt, against the background of the practice reconstructed by Andreozzi and the notes on Falloppia's statements that have come down to us from the pens of various students, that the young anatomist himself gave the lethal opium to those condemned to death. Maffei, the translator of the passage in the Italian edition of Falloppia's surgical works also understood that it was Falloppia personally who gave the deadly opium. He translated that the anatomists were given a criminal ("malfattore") so "we would make him die of whichever death we deemed convenient" ("perche lo facessimo morire di qual morte a noi pareua conveniente") and to dissect him. The translator quotes Falloppia as saying, in the first person singular, "I gave two drachms of opium" and "I gave it again."[84]

The advantages of obtaining bodies for anatomy in this manner were obvious. Since the timing of the death was in the anatomist's hands, the cadavers would be very "fresh" and, unlike those of decapitated or hanged men, all anatomical structures would remain perfectly intact. For modern sensibilities – as already for those of Jean Astruc – it is of course unimaginable and seems in the highest degree reprehensible in moral-ethical terms that an anatomist kills with his own hands the people he wants to dissect afterward. However, Falloppia's repeated statements on the subject, as documented also in the personal, unpublished notes of his students, do not indicate that he considered such actions wrongful or felt guilty. Nobody forced him to keep telling his students about it and there is not even a hint of an apology. Clearly, he saw no reason to hide what he had done.

If we want to arrive at an adequate historical understanding, we have to adopt the contemporary perspective for a moment. From this perspective, it becomes comprehensible why Falloppia, by all appearances, did not feel guilty of any wrongdoing. From a contemporary viewpoint, these men and women had forfeited their lives. They were certain to die, paying for their crimes. In the sixteenth century, high-ranking princes such as Archduke Ferdinand II, the French king, and the Pope even supported or initiated experiments that were designed to test the efficacy of antidotes against deadly poisons on men and women who had been condemned to death.[85] In the worst case, the experiments ended in a protracted and excruciatingly painful death, when for example corrosive sublimate of mercury was given. On the other hand, the victims and their relatives were spared the shame of a public execution. Even more importantly, if they were lucky – and various such cases are documented – they survived the trial and were pardoned or at least escaped capital punishment. The request of the man with quartan fever that Falloppia give him a second dose of opium probably has to be seen in this light, too. It is therefore quite possible that the physicians and others responsible for these trials spoke the truth when they claimed that their "test persons" had expressly agreed to such experiments. And since the poison had to be taken by mouth, some degree of cooperation was virtually indispensable anyway.

If trials that involved giving a deadly poisons and then an antidote to people who had been sentenced to death were acceptable even to the Pope, killing criminals with a deadly dose of opium who were sure to be hanged or decapitated otherwise could seem almost like a privilege from a contemporary perspective. They and their families were not only spared the shame and humiliation of a public execution. They were given the kind of death many people desired, once the end was imminent and inevitable. The famous Spanish Juan Fragoso (ca 1530–1597) in his *Cirugia universal* explicitly praised the Italian practice, where, in his words, the anatomists, at given times of the year, asked the judges for some men who had been condemned to death and gave them wine to drink with two or three drachms of opium. When they had drunk the wine, they were, at first, happy and in good spirits, he claimed, and then, when the opium reached the vital organs and the heart, they fell into a profound sleep.[86]

Falloppia's findings and discoveries

To this day, Falloppia is considered one of the leading figures in the history of anatomy. He published most of his findings in his *Observationes anatomicae ad Petrum Mannam,* which came out in 1561 with Marcantonio Olmo in Venice.[87] As Falloppia explained in his letter to the reader, he had written the book already four years earlier, at the request of the Cremonese physician and anatomist Petrus Manna.[88] He did not explain why he initially refrained from publishing it and he clearly did not yet put the final touches in 1557. The book refers among others to Realdo Colombo's *De re medica,* which came out in 1559 only. In 1562, further editions of the *Observationes* appeared in Cologne and Paris.[89] In 1588, Johannes Siegfried brought out a revised edition in which he divided the material into five books.[90] Of course, the work was also included in the various editions of Falloppia's *Opera.*

The *Observationes anatomicae* was conceived as a long epistolary treatise to Manna. Little is known about Manna, except that he was a personal physician to Christina of Denmark, the wife of Francesco II Sforza. Colombo probably was referring to him when he mentioned a "Petro Manaae" as one of the physicians who saw him do dissections in Cremona and asked him to perform the vivisection of a dog.[91] Presumably this was before Colombo left Cremona in 1535, which suggests that Manna was already a practicing physician at that time.[92] A portrait of Manna by the painter Lucia Anguissola shows him in 1557 at a settled age, perhaps around fifty-five to sixty years old, which would indicate that he was born around 1500.[93] His image has also survived in Cremona on a contemporary medal, with the inscription "PETRVS MANNA MEDICVS CREMONENSIS".[94] Archival documents make it possible to narrow down the time of his death to the period between 1558 and August 1560, by which date he was dead. So, when Falloppia's *Observationes* came out, Manna was no longer alive.[95]

The *Observationes* was a very original work. There was nothing comparable to this "excellent work" ("eximium opus"), Albrecht von Haller praised it in his *Bibliotheca anatomica.*[96] As Falloppia explained in his letter to the reader, his intention was not to offer a systematic outline of human anatomy and the book certainly was not designed to serve as an introductory work for students.[97] The focus was decidedly on the new and unknown. Falloppia discussed, one after the other, numerous muscles, bones, and other anatomical structures, contradicting or expanding on the descriptions given by Vesalius, Galen, and other anatomists. The principal merit of Vesalius' anatomical research lay in the fact that he systematically reviewed and, if necessary, corrected the descriptions of anatomical structures which Galen had given based on his dissections of animals and which had been erroneously accepted as valid descriptions of human anatomy. His iconic status has somewhat obscured the fact, however, that Vesalius did not make any important new discoveries of his own. Moreover, he himself committed serious errors, which rather ironically resulted, in particular, from his own reliance on the dissection of animals.[98] Falloppia played in a different league here. He discovered a range of new unknown anatomical structures and described them for the first time ever.

Falloppia clearly was proud of his findings. In this, he was a characteristic representative of the rise of a new phenomenon that has stayed with us and is very familiar today, namely priority disputes. As Falloppia explained, the *Observationes* was also directed against those who claimed to have made anatomical discoveries, which they actually had from him or from students who had attended his anatomical demonstrations. On top of that, they sometimes reproduced his observations in a "distorted and corrupt" way. Even some of his beloved pupils, whom he had instructed with the greatest care in the art of anatomy, were guilty of that. In order to give authority and weight to their own words, they contradicted Falloppia and sometimes reproduced his statements in such a twisted way ("distortas et falsas") as he could not even dream of. His book would protect his readers from such fraud. When he mentioned new anatomical findings, he would state who was the true discoverer.[99] Undoubtedly Falloppia was above all referring to himself here but also to colleagues like Giovanni Filippo Ingrassia (1510–1580) in Naples, whose discovery of the stapes, the hitherto unknown third auditory ossicle, was falsely claimed by Realdo Colombo, as was the discovery of the clitoris.[100]

In the *Observationes*, Falloppia, then in his late thirties, announced another, future work, which would offer a comprehensive account.[101] In this *volumen anatomicum*, he would cover the entire anatomy down to the smallest detail. It was to be adorned with numerous illustrations, which would show the individual parts also in their respective anatomical context, with the structures to which they were connected and which surrounded them. In other words, his intention – this is how I would interpret this announcement – was to depict not only the different bones, muscles, tendons, nerves, and vessels in the body in isolation but also in situ. He would show not only the anatomy of man, he added, but also that of the monkey, so that the teachings of Galen, whose anatomy was largely based on the dissection of animals, would be more easily understood.[102]

This book would undoubtedly have become one of the major works of Renaissance anatomy and, if I understand Falloppia's concept correctly, the first ever illustrated atlas of topographical anatomy in the history of medicine. Falloppia did not live to complete it and we do not know how far Falloppia's work on this comprehensive, illustrated textbook had actually progressed. After Falloppia's death, the Duke of Este immediately asked that manuscripts penned by Falloppia and books containing his handwritten annotations be searched for. He could assume that Falloppia, like most scholars of the time, had left behind a more or less extensive handwritten *nachlass* and apparently he hoped to be able to acquire it.[103] However, Falloppia's manuscripts and books were never found. Perhaps Melchior Wieland, who had lived in the same house with Falloppia for years, took them with him when he moved into another house, near the botanical garden, after Falloppia's death.[104] At any rate, to this day, not a single manuscript and not even a single book from Falloppia's possession has surfaced.[105]

The *Observationes* are the major source for Falloppia's anatomical findings. In addition, I will draw on the student notes of Georg Handsch and the

Helmstedt Anonymus. They were not designed for the book market but for personal use only and thus can be taken to render Falloppia's teaching as faithfully as the students could. Moreover, they reflect Falloppia's anatomical teaching shortly after his arrival in Padua and sometimes document anatomical findings that Falloppia presented in his lectures and demonstrations already several years before the *Observationes* came out and thus can help us trace more precisely when and how Falloppia arrived at certain insights over the years, for example, about the ovarian ducts and the lacteals.

Some student notes on new anatomical findings have also come down to us in print. In 1570, Franciscus Michinus published his notes Falloppia's lecture on Galen's *De ossibus*, supplementing them with a handful of *Observationes anathomicae* [*sic*!] on specific structures that Falloppia had shown to his students.[106]

The brothers Meietus in Venice moreover published a survey of human anatomy, in 1571, under the title *De humani corporis compendium*.[107] In 1585, a second edition came out[108] and it was later also published with in Falloppia's *Opera*, now under the title *Institutiones anatomiae*. In a brief introductory letter, the unknown scribe explained to an unnamed friend that he wanted to summarize the "humanae fabricae historiam" in a short compendium. In particular, his aim was to point out all the things Falloppia had taught against the *dogmata* of other anatomists. Historians have paid little attention to this work.[109] Albrecht von Haller already described it as written by some ignorant disciple of Falloppia.[110] Undoubtedly, the *Observationes* is a work of much superior quality. As a source for the development of Falloppia's insights, the *Compendium* is useful, however. A comparison with the notes of the Helmstedt Anonymus suggests that the printed text – which is not quite so short after all, with about 170 pages in the 1571 edition – was indeed based on student notes. Not only did the scribe render Falloppia's words in the first person ("enarraverimus", "enarrabimus", etc.).[111] Also in terms of content, the compendium shows many similarities with the notes of the Helmstedt Anonymus, which cannot be explained simply by the fact that different verbal descriptions of one and the same body part inevitably use somewhat similar terms. The central literary references are the same, such as Galen's *De dissectione musculorum*, and, above all, the numerous passages in which the scribe – like the Helmstedt Anonymus – quite bluntly refers to "errores" of Vesalius are striking. We do not know when the student notes on which the *Compendium* was based were originally taken but it was very likely in the mid-1550s. As we will see, the parts on the female reproductive organs and on the *tubae uterinae* document insights that went beyond what we find in the manuscript student notes of Handsch and the Helmstedt Anonymus from the early 1550s. However, the description was not yet as precise as that in the *Observationes*.

The number of anatomical structures, which Falloppia discovered or which he described at least far more precisely than previous authors, is long and impressive. Some authors have compiled long lists or indeed provided detailed accounts of Falloppia's many findings.[112] Others published articles specifically on how he described specific organs and parts of the body, such as the cranial nerves, the eye muscles, or the female genitals. In what follows, I will look at a fair range of Falloppia's more important findings but there are many others, which I will, at best, only mention briefly, all the more so when they have already been studied in detail by other historians or by experienced anatomists. Although Falloppia is well known, for example, for having provided the first clear description of primary dentition, of the follicle of the tooth bud, and of how the first teeth are eventually replaced by the permanent teeth, I will not devote a special subchapter to his work on the teeth.[113]

Female genitals

The most famous findings which Falloppia published in his *Observationes* are undoubtedly those referring to the *tubae uterinae*, the fallopian tubes as they are commonly named after him today. They extend on both sides from the upper part of the uterus and end freely in the abdominal cavity, near the ovaries. According to our modern understanding, the female ova from the ovaries enter these tubes and if conception takes place this is where the sperm usually encounters the egg. The fertilized egg then moves slowly through the tube to the cavity of the uterus and is implanted there.

Since antiquity, Herophilos, Galen, and many other authors up to and including Vesalius had described "seminal ducts," which, they believed, carried female semen from the female testes to the uterus. They conceived them as analogous to the seminal ducts in the male.[114] In his demonstration of the uterus of a ewe, in January 1552, which was documented by both the Helmstedt Anonymus and Georg Handsch,[115] Falloppia, like many other anatomists in this period, declared the old idea as ridiculous that men and women had the same genitals, which in the male were only turned inside out.[116] He still adhered to the traditional Galenic account of the uterus, however, which had only recently been confirmed by Vesalius. Falloppia described in detail the (animal) uterus with its "horns" and its shape, which was similar to that of a calf's head. In animals – but not in humans – it was divided into two cavities ("sinus"). He distinguished the body of the uterus from the *os matricis* (the cervix of modern terminology) and the *cervix* or *collum* (the vagina). The ovarian ducts are only vaguely hinted at: Falloppia told his students of the female testes, with their arteries and veins, and of short *processus varicosos* which contained the seed and emitted it into the uterus.[117]

His anatomical work in the following years led Falloppia to a major revision. According to the *Compendium*, Falloppia explained to his students that he had set out to examine more closely those vessels, which were commonly believed to allow the female semen to pass from the female *testes* to the uterus. These vessels, Falloppia now explained, extended from the uterus to the *testes* but were not connected with them. They ended in the surrounding cavity. Widening at the end, they formed a kind of contorted vesicle, similar to those vessels at the neck of the bladder which in the male contained the semen.

They were permeable. A probe could be advanced to the uterus without damaging anything. In Falloppia's estimation, these passages allowed the uterus to free itself of smoke as if through a chimney. Falloppia was not sure, however, whether this was their primary purpose ("usus"). In any case, it seemed to him that the *testes* – which were thus not directly connected to the uterus – did not serve to produce semen in the woman. He intended to gain more precise information by more frequent and more careful dissections.[118]

As outlined above, the *Compendium* was published in 1571 but was clearly based on the notes taken by one of his students some years before the publication of the *Observationes*, probably around 1555. In terms of content, the description of the *tubae uterinae* in the *Compendium* offers an intermediate step, in fact, toward Falloppia's final account in the *Observationes anatomicae* of 1561. In the *Observationes*, Falloppia presented the results of his further research, based on the dissection of humans, ewes, cows, and other mammals. He first described the female *testes*. Everybody claimed that they were filled with semen but despite his many efforts, he had not been able to discover anything of the sort. He found them to contain vesicular structures that were turgid ("turgentes") with a sometimes yellowish, sometimes translucent watery fluid. In retrospect, this was the first precise description of the structures we now call follicles. The female seed by contrast, Falloppia proclaimed, could only be found in the lumen of the ducts; presumably, he based his conclusions on some fluid matter he found there.[119] Again he firmly rejected the traditional claim that these "meatus seminarii" in women began in the testes and were directly connected to them and led to the horns of the uterus. He had never found that these ducts started directly at the testes and were connected with them, except in diseases. They ended at some distance from the testes, about half a finger's width away, and were not connected to them by any vessel. All that could be found was a very fine *membrana peritonealis*, which enveloped the *testes* as well as their vessels and the *vas deferens*.[120]

Falloppia still did not commit himself to a specific understanding of the function of the ducts. His reference to the semen which he had found in them suggests that he believed it was formed in the area of the fimbriae but he did not discuss this issue. He concentrated on the morphological description. The *meatus* was fine and fibrous, initially sinewy and pale at the uterine side but it widened as it progressed toward the *testes*. Toward the end, its tortuous folds straightened. It became quite wide and seemed membranous and fleshy because of its red color at the end. Its ending, with what we call *fimbriae* today, was irregular and fringed, like the edges of worn cloth. When the fringes were out of the way or removed, a wide opening appeared. This widening inspired Falloppia to name the duct "tuba". The English translation as "tube" or "duct" does not adequately render the original meaning of the term. The "tuba" was a wind instrument, a kind of trumpet, which the Romans used to send signals in their military campaigns. If we were to take Falloppia by the word, we would have to speak of uterine "trumpets", in fact, rather than of "tubes". Apart from the similarity in shape, he may also have had in mind the explanation of its possible use that we find in the *Compendium* only, namely that they served as a kind of chimney and allowed the uterus to "blow out" the vapors contained in it.[121]

With regard to the vagina and the cervix in modern terminology, Falloppia criticized the modern anatomists who almost unanimously called the part into which the penis entered in intercourse the "cervix".[122] Galen and Soran had rightly called it "aidoion gynaikeion" or "kolpos gynaikeios", not "cervix". The modern anatomists were ignorant of the "true cervix" ("veram ceruicem"). This was the much smaller structure with a narrow opening orifice ("ostiolum angustum"), at the very end of the "kolpos" (Falloppia did not use the word "vagina" here). The penis only touched it but could not enter it. Only the semen passed through the opening.[123]

In his lectures, Falloppia also described as an error the widespread belief that the cervix closed tightly after conception, so that not even a thin pin could pass through. It was true that the neck of the uterus contracted after intercourse and was like glued together so that the semen could not escape.[124] But it remained permeable. Falloppia had tried this in pregnant dogs with a probe. This also explained why some women could still evacuate menstrual blood at the beginning of pregnancy.[125]

As to the hymen, Falloppia changed his opinion over the years. In the early 1550s, he still denied the existence of a hymen in virgins that ruptured at first intercourse, as Berengario da Carpi and others had claimed.[126] This was a "fable" ("fabula") he thought. He had dissected three virgins and found nothing of the sort. If blood showed after the first intercourse, it was only because the virgin's vagina was narrow and not yet used to intercourse, so that some veins ruptured.[127] By the time he wrote his *Observationes*, he had changed his mind. Now he opposed those anatomists who ridiculed the belief in a hymen. According to Falloppia, a fibrous membrane could be found in the vagina of virgins, just above the orifice of the urethra. It had a hole in the center, which in adult women easily accommodated the tip of the little finger. When it was stretched and torn during the first intercourse, it hurt.[128]

In historical literature, Falloppia has sometimes been praised for his discovery of the clitoris – or criticized for falsely claiming this discovery, like Colombo,[129] for himself.[130] A closer look at the *Observationes anatomy* proves both assertions wrong. Falloppia carefully described the clitoris, the "bifurcation" of two *crura*, which joined at the tip, the vessels that on back, the spongy structure in side, and the skin that covered the tip or *glans* like a foreskin.[131] He explicitly quoted Avicenna and Albucasis, however, who had already mentioned this structure, which was said to sometimes grow so large that women could have sexual intercourse with each other. "Virga" – like the male member – or "albathara" Avicenna had called it, "tentigo" it was called in the Latin translation of the work of Albucasis. The Greeks, according to Falloppia, however, called this part "klitorida".[132] Criticizing the anatomists of his time ("anatomici nostri"), who did not mention this little, hidden part, he praised himself as the first one who had demonstrated it in recent times ("vt ego primus fuerim, qui superioribus annis idem pate-fecerim"). If others – presumably he was referring to Colombo – talked about it, they owed their knowledge to him or his students. In other words, Falloppia did not style himself as the very first anatomist who described the clitoris but as the first among the moderns who had rediscovered it.[133] This more limited claim was probably justified. The *Observationes* came out after Colombo's *Re anatomica* but in the *Compendium*, which by all appearances documents Falloppia's lectures – and his more limited stated of knowledge – in the mid-1550s, the clitoris is already mentioned briefly, though without giving it a name. The *Compendium* talks of a protruding body in the upper part of the *pudendum*, which in some women was so long that it looked like a penis, especially when the clothes rubbed on it and made it inflate and become turgid in the way of a penis.[134]

The (re)discovery of the clitoris in learned medicine had far-reaching consequences for the perception of female sexuality and the female body. In accordance with the teleological concepts of Renaissance Galenists, it implied that Nature or God had equipped women, on purpose, with a part that was designed to give them sexual pleasure. In this respect, Colombo had indeed been more explicit than Falloppia. In other contexts, for example, when it came to the size of the penis, Falloppia underscored the importance of female sexual pleasure, which prompted the emission of female semen and thus was essential for the procreation of the human species.[135] In his account of the clitoris, he only mentioned the similarities with the penis, however, thus associating the clitoris directly with sexual intercourse but by implication only with sexual pleasure.

The very explicit comparison with the penis had a profound effect, in turn, on the way learned medicine perceived the female body. Falloppia was one of the leading authors who – pace Laqueur – stressed the many fundamental differences between the male and the female genitals rather than assuming that one was basically an extroverted version of the other. The Galenic idea of a basic analogy remained alive, however. Now, the rediscovery of the clitoris challenged even this idea of a mere analogy. The clitoris and the vagina could not both correspond to the penis. If the clitoris was a natural part of female anatomy and not a pathological excrescence, women would have had, like hermaphrodites, "male" and "female" genitals at the same time.[136]

Ileocaecal valve

As the student notes show, Falloppia also described in detail – and thus, according to all we know, for the first time in history – the ileocaecal valve which was later named after Caspar Bauhin. It is a functional barrier formed by skin folds between the small intestine and the large intestine, the colon. It allows digestive pulp to enter the colon from the small intestine but prevents fecal matter from flowing backward into it. Handsch documented

Falloppia's description of this "valve" and his explanation of its function in considerable detail on an inserted slip of paper, probably in 1552.[137] As Falloppia explained, he had found that "water, when poured into the small intestine, flows through the whole intestine; when poured into the intestine through the rectum, it does not flow through the whole intestine but is stopped and collects at this thick beginning of the colon". The *caecum intestinum* was attached in such a way as to close the entrance to the small intestine when it filled. From this, according to Falloppia, derived also the function, "namely, that the feces do not regurgitate to the higher parts". Nature used the same kind of device, he explained, to keep the urine in the bladder from flowing back into the ureters.[138]

In early February 1553, Falloppia seems to have demonstrated the *caecum* and the valvular structure again in the private anatomy of a monkey, in which both Handsch and the Helmstedt Anonymus participated. He described the function of the valve and now compared it with that of a heart valve. "Where the caecum attaches to the ileum", the Helmstedt Anonymus noted, "there are two folds which are compressed when filled or inflated, as happens in the heart, and which prevent the movement back [into the ileum]." This was shown by the fact "that water given into the rectum, or a wind that enters the caecum [from the rectum] cannot pass through from the colon; and when it [the water] is put in from above, it passes through".[139] Handsch also noted the prevention of a reflux of fecal matter and described how Falloppia first filled water and then air into the intestine. When he poured the water into the small intestine, it ultimately flowed into the rectum. When he did so from the colon, only the caecum filled. Not even the air could pass through toward the small intestine.[140]

The finding had important implications for a widely used therapeutic procedure, namely the administration of medicine via an enema. Since neither fluid nor air could pass from the caecum into the small intestine, the medicines in the enema could not have the desired effects on the intestine, leave alone the stomach. Some authors believed they had witnessed sick people vomiting up the enema fluid but, according to Falloppia, this could happen only exceptionally, when the intestines were weakened by disease.[141]

Lacteals

The lacteals, as we understand them today, are lymphatic vessels that originate from the small intestines, where they absorb the fatty substances from the digested food and feed them as liquid chyle into the larger lymphatic vessels. Together with the lymph of the lower extremities and the abdomen this chyle ultimately enters the venous blood via the *ductus thoracicus* at the junction of the left subclavian and the left internal jugular vein. Most of the time, the lacteals are difficult to see with the naked eye but they fill with a whitish-yellowish fluid shortly after food has been taken in. They owe the name "lacteals" to this "milky" color.

The first references to the lacteals can be found in ancient literature. Galen had already described vessels in the abdomen, which contained a turbid fluid rather than blood.[142] Today, Gaspare Aselli (ca 1581–1626) is generally regarded as the true discoverer of lacteals. He described them, based on accidental observations during the vivisection of dogs, in his (posthumously published) *De lactibus sive lacteis venis* (1627).[143] However, already about seventy years earlier – probably in 1552 – Falloppia, according to Handsch's notes, showed his students "ducts" ("meatus") in the abdominal cavity of a dog, which, he thought, neither Galen nor Vesalius had mentioned, and which, according to him, also existed in humans. To a student who continued to take notes – presumably, most students were standing around the dissection table and looking rather than writing – Falloppia dictated a more detailed description: the ducts had numerous branches and extended between the lower part of the liver, on the side of the portal vein, and terminated in the deeper layer of the omentum and in the pancreas. They contained a thin, yellow fluid, that had no particular taste or was, at most, bitter. Their function, he confessed, he had not yet been able to discern; perhaps they served to nourish the pancreas.[144]

A somewhat more precise description was published by Francesco Michino in 1570, in his notes on anatomical observations made during Falloppia's dissections in 1554.[145] Falloppia now explicitly described the contents as "oily" and also referred to the "glandulae" – presumably, the lymph glands of modern terminology. He continued to assume a relation to the pancreas: "On the lower side of the liver are certain small ducts which end and stop in the pancreas and in the nearby glands, and these ducts transport a yellow, oily juice which tends somewhat to bitterness."[146]

Like Eustachio's discovery of the thoracic duct, at around the same time,[147] Falloppia's discovery went largely unnoticed in the decades after his death – until Aselli rediscovered the lacteals.

Brain and cranial nerves

Another anatomical region where Falloppia made significant new discoveries and offered more precise descriptions of the known structures was the base of the skull and the cranial nerves. Anatomical knowledge of the complex structure of the human brain and the cranial nerves advanced considerably in the sixteenth century. Major pioneering work was done by Berengario da Carpi.[148] His account of the cranial nerves is more precise and complete than that of Galen and, in fact, also that of Vesalius, whose passages on the cranial nerves O'Malley counted among the "least important sections" of the *Fabrica*.[149]

The notes of Georg Handsch and the Helmstedt Anonymus do not offer a systematic account of Falloppia's demonstration and explanation of the cranial nerves which might allow a more precise dating of the findings he presented in the *Observationes*. Moreover, the anatomy of the cranial nerves is particularly difficult and complex and very few readers will be familiar with it. I will limit myself here to a brief summary of Falloppia's account, directing readers who desire a more detailed analysis to O'Malley's study of the relevant section in the *Observationes*.[150]

Falloppia followed the old method, canonized by Galen, of giving numbers to the individual cranial nerves. This was common at the time also for muscles but in the case of the cranial nerves, it has survived until today, with the numbering of the cranial nerves from I to XII. Following his account of the first, optical, and the second, olfactory, nerve, Falloppia introduced considerable changes however. In the case of the *nervus trigeminus* of the modern terminology, whose branches extend to the eyes, the nose, and to the upper jaw and lower jaw, he described the nerves, which were previously considered to be the third and fourth pair as a single pair of nerves. In his view, the decisive factor for the distinction and numbering of individual nerves was not the number of openings in the base of the brain or even the number of branches. One cranial nerve could exit through several openings and several different cranial nerves could use the same exit. The primary criterion was whether two nerves exited the dural membrane separately or together.[151] As a fourth pair of cranial nerves, Falloppia now identified the *nervus abducens*, which innervates the corresponding eye muscles.[152] The fifth pair, following the traditional counting, comprised what we know today as the *nervus facialis* and the *nervus acusticus*. Falloppia considered the *nervus facialis* a separate nerve but did not want to give it a number of its own, to avoid constantly contradicting the anatomists before him, as he explained.[153] In this context, he also described the winding osseous canal through which the facial nerve passes through the temporal bone. Today this *canalis nervi facialis* is also known as the "Fallopian canal" or "Fallopian aqueduct".[154] He also maintained the traditional count for the sixth pair of nerves but distinguished the *nervus glosso-pharyngeus* of the modern terminology, from the *nervus vagus* and the *nervus accessorius*.[155] He had nothing new to add to the account of what we call the hypoglossal nerve today, which supplies the muscles of the larynx, the hyoid bone, and the *processus styloideus*.[156] There was one more, eighth pair of nerves, he concluded his account, however. It was the nerve we know today as the trochlear nerve. Achillini had already vaguely described it in 1520.[157] Vesalius had traced its peripheral course but assigned it to the third cranial nerve. Falloppia now isolated it in its entirety and showed that it exclusively supplied the oblique superior eye muscle, whose tendon runs over the *trochlea*.[158]

Eye muscles

Falloppia devoted considerable space in his *Observationes anatomicae* to the muscles of the eye, some of which are quite fine and difficult to isolate.[159] In two central points, he corrected and improved on extant anatomical research: he denied the existence, in humans, of the muscle, which we call today *musculus retractor bulbi* and he described for the first time the *musculus levator palpebrae*.

The first point offered a particularly striking example of a serious methodological error on the part of Vesalius. Vesalius' central criticism of Galen's anatomical writings was that he had largely relied only on dissections of animals. In the case of the eye muscles, Falloppia caught Vesalius himself presenting findings as human anatomy, which he had only seen in animals. In addition to the six muscles that accomplished the movement of the eyeball upward, downward, sideways, and in a circle, Vesalius listed a seventh muscle that anchored the eyeball in the orbit and pulled it toward it. He described its insertion and shape and showed it on the corresponding illustration of the eye muscles.[160] This muscle can only be found in animals, however, and with his illustration in a textbook on human anatomy, he provided fake visual evidence for this error. Falloppia refrained from open criticism of Vesalius on this point. However, already in 1551/52, several years before Realdo Colombo made his own doubts public in 1559,[161] Falloppia explained to his students that this muscle existed only in animals. Dissecting the eyes of monkeys and cattle in a private anatomy, he showed this seventh muscle, adding that it occurred only in cattle, dogs, and other animals but not in humans.[162]

Falloppia made his second major discovery on the eye muscles, the *musculus levator palpebrae* of modern terminology, shortly after. In 1550/51, Fracanzano still explained to his students that the organ which made it possible to raise the upper eyelid was not known. The movement could not originate from the eye itself and the *spiritus* that arrived there. He had seen a soldier whose eye was completely hollowed out but who could still lift the upper lids.[163] In January 1552, Falloppia had no answer either. He had searched for the tools that made this movement possible but had not been able to find them. He rejected the theory that the muscles that were responsible for other movements of the eye somehow worked together. Circular muscles could not achieve a straight movement. He assumed that there were fibers in the upper eyelid that were responsible for this movement but could not be separated and isolated.[164] As Falloppia reported in his *Observationes anatomicae*, the dissection of an animal eye then put him on the right track. In late December of 1552, a certain Matthias Guttich brought him the head of a dolphin ("phoca") from the fish market in Venice.[165] Falloppia dissected the eye and found this hitherto unknown muscle responsible for lifting the upper eyelid. He showed this muscle in a public anatomy, in January 1553.[166] After this discovery, he searched for the same muscle in the human eye and found a small and very slender muscle that fanned out into the upper eyelid.

Falloppia also described other subtle structures in this area, such as the tendon of the fifth eye muscle – the *musculus obliquus superior* in modern terminology – which changes its direction thanks to a pulley-like structure, the *trochlea*.[167]

Auditory organs

One of the great advances of Renaissance anatomy was the first precise account of the small auditory ossicles that pick up and amplify the vibrations of the eardrum and, according to modern understanding, help convert them into nervous impulses.[168] Falloppia made major contributions here.[169] Handsch once again supplemented his notes on the anatomical demonstration by Fracanzano, which he had attended in the winter of 1550/51, with what he learned from Falloppia. Falloppia, he noted, showed them, in the head of a calf, the "membranula" at the end of the auditory canal and two ossicles, the hammer and the anvil; the notes do not mention the stapes. The ossicles were inside a cavity, which, he explained, was filled with air from birth. When the "membranula" was set in motion by the effect of the sound on the surrounding air, the air inside the cavity was set into motion as well and the sound was transmitted to the *sensus communis*. Falloppia also mentioned a test that is still used today for distinguishing between sensorineural hearing loss and impaired sound conduction due, for example, to a blocked external ear tube; in the latter hearing is still preserved via bone conduction. It was a good sign, he explained to his listeners when someone hard of hearing could still hear noise when he put a finger in his ear.[170]

Just a year later, watching Falloppia's dissection of a monkey in February 1553, the Helmstedt Anonymus also took note of the third auditory ossicle, the stapes ("ossiculum stapes dictum"). It is the earliest known document that mentions it in writing.[171]

Falloppia did not claim the discovery for himself. In the *Observationes*, he explained how he found out about the auditory ossicles.[172] Judging from their writings, they were not known to the ancient authors. It was Berengario da Carpi who finally described the hammer and the anvil.[173] The stapes, however, had been discovered neither by Vesalius nor by Realdo Colombo but by Giovanni Filippo Ingrassia when he taught anatomy in Naples. Falloppia told the story how he heard about this discovery. When he started teaching anatomy in Pisa in 1548, a listener, who was apparently a relative of Ingrassia's, came to him and told him that the latter had found a third ossicle in the tympanic cavity ("tympanum"), which he called "stapes" (i.e. stirrup) because of its shape. Falloppia then searched specifically for this ossicle, found it and showed it publicly, to the great astonishment of those present. He also described the stapes in his letters, including one to friends in Rome. As they reported to him, neither Colombo, who was now teaching in Rome, nor anyone else there had heard them tell of the stapes. Apart from Ingrassia and Colombo, there was no other suitably skilled teacher of anatomy at the time, except Giovanni Battista Canani, with whom Falloppia remained in contact and who he had not mentioned the stapes either. So the credit for the discovery undoubtedly went to Ingrassia.

Falloppia had good reasons for telling this story in such detail: Realdo Colombo, who like Pietro Manna, the addressee of Falloppia's *Observationes*, was from Cremona, had claimed the discovery of the stapes in his *De re anatomica* of 1559. The newer anatomists, he explained, had known two auditory ossicles since Berengario da Carpi. He had found a third one on close examination, which resembled a stirrup with its shape and its hole in the middle.[174] Falloppia made it clear that Colombo – who had taught before himself in Pisa, where apparently no one had heard of the stapes by 1548 – was wrongly claiming this discovery for himself. He suggested that Colombo had learned from others about Ingrassia's discovery or heard about Falloppia's demonstration of the third auditory ossicle, to which Colombo moreover gave exactly the same name "stapes" that Ingrassia and Falloppia had used years before Colombo's work came out.[175]

Heart and major vessels

In the anatomy of the heart and the major vessels, Falloppia's primarily contributed to a better understanding of the fetal vessels.[176] He discovered – or rather rediscovered – the *ductus arteriosus*, a short vascular shunt through which, according to our modern understanding, part of the blood from the fetal pulmonary artery flows directly into the aorta rather than passing through the lungs. It usually closes after birth, leaving the *ligamentum arteriosum* behind. The discovery has been widely ascribed to Leonardo Botalli (1519–1587) and is known today also as the *ductus arteriosus Botalli*; its discovery has also been attributed to Giulio Cesare Aranzi (1530–1589). When these two published their discoveries, however, Falloppia had already provided an account of the *ductus arteriosus* in his *Observationes anatomicae*.[177] He did not claim the discovery for himself: Galen, he declared, had already mentioned it, though with only a few words.[178] Without explicitly naming Vesalius, he expressed his astonishment, however, that almost all anatomists had been negligent and overlooked this artery or canal ("canalis"). The blood flowed through it in utero and later it dried up and was so thick had it could not escape the senses.[179]

Some historians have also credited Falloppia with a major role in the discovery of the venous valves. His student and successor Fabrizi d'Acquapendente later described them in detail but did not recognize their function. The venous valves obstruct the flow of blood from the center into the periphery and this made little sense within the framework of Galenist physiology, which maintained that the natural movement of the blood in veins was from the center toward the peripher toward the extremities and other parts of the body, to nourish them. The venous valves eventually provided William Harvey with an important building block for his revolutionary theory of blood circulation, however.

Falloppia was one of the first authors who explicitly mentioned the venous valves. In his *Observationes*, he quoted the report of João Rodrigues de

Castelo Branco called Amatus Lusitanus (whom some Portuguese authors would like to see as the true discoverer).[180] Amatus had reported that a little flap ("ostiola") had been found in the *vena azygos* in twelve dissections of animals and humans, which were performed in 1547 in Ferrara. According to Amatus, Giovanni Battista Canani and other learned men had seen it. Amatus described moreover an experiment as proof that it really functioned as a valve. If one inserted a little tube into the cut-off *vena cava* and blew air into the vein through it, he claimed, the *vena azygos* would fill with air. But if one put that tube into the azygos vein below the *ostiola* and blew into it, the air did not reach the *vena cava*. The valve, he concluded, prevented blood from flowing back from the *vena azygos* into the *vena cava*, much as the valve at the entrance of the ureter into the bladder and the heart valves prevented a reflux. This was important in the treatment of pleurisy. If, in the experience of many physicians, bloodletting on the same arm was helpful in pleurisy, the reason could not be that the morbid matter was drawn from the area of the *vena azygos* via the *vena cava*. Another explanation was needed.[181]

Falloppia firmly rejected Amatus' claim. He had never seen such a valve at the junction of the azygos vein in his dissections. Nor did he want to believe that Canani, that eminent anatomist, had claimed its existence. At most, he might have been jesting, joking with some of Amatus' companions.[182] In his *Examen* of Falloppia's *Observationes*, Vesalius, for his part, explained that Canani himself, years ago, when they were treating Francesco d'Este together at the Imperial Diet in Regensburg – it must have been the one in 1546 – had told him of his discovery that at the beginning of the *vena azygos*, of the *venae renales*, and in a vein near the *os sacrum* there were valves or "membranae" whose function was similar to that of the valves at the beginning of the *arteria pulmonalis* and the *aorta*.[183]

Amatus' account was misleading and his claim that this flap had been seen in all twelve dissections in 1547 was, in retrospective judgment, in all probability simply invented. From today's perspective, we would expect that the experiment he described would prove exactly the opposite of what he claimed: the air – and thus also the blood – would have to be able to pass only from the *vena azygos* into the *vena cava* and from there to the heart, not in the opposite direction. Amatus described the outcome in line with what Galenist physicians would expect, who thought that the blood in the veins moved toward the parts to nourish them rather than back to the heart or the liver. Amatus moreover described the valve in the *vena azygos* as if it were located at the beginning of the vessel, like the heart valves. There is no such valve at the confluence (or, from the perspective of Renaissance physicians, the beginning) of the *vena azygos*. The vein widens initially. A valve is encountered only a little further away from the *vena cava* and is not always found either. It is quite possible that Amatus had not seen this flap himself and misunderstood what others had told him about it. What is surprising, is that Falloppia, despite his close association with Canani, apparently heard nothing of these flaps during his own time in Ferrara. In the records of Falloppia's Paduan students, only the cardiac valves are mentioned.

Based on the accounts of Amatus and Vesalius, it is widely accepted today in historical research that Canani saw the venous valves in the human body no later than 1546 and thus well before Fabrizi.[184] However, to what extent he may be considered as their true discoverer is uncertain. As Albrecht von Haller already pointed out in 1751,[185] French anatomists had made corresponding observations even earlier. In 1545, Charles Estienne published in Paris a description of certain "exortus", "apophyses", or "epiphyses", that is, "excrescences" of the membranes in the liver, to which he attributed the function of preventing a reflux of the blood into the liver.[186] Albrecht von Haller praised Estienne as the first describer of venous valves. I find it questionable, however, whether Estienne truly saw venous valves in the modern sense. He did not describe these membranes in the large veins outside the liver but in the parenchyma of the liver itself. If one incised it with a scalpel, Estienne said elsewhere, one saw blood emerging from the smallest openings, and one also saw "epiphyses" of membranes, which, he assumed, prevented the blood from escaping to the outside, so that it remained longer in the liver and was all the better concocted.[187]

By contrast, in the chapter *De membranis* of his *In Hippocratis et Galeni physiologiae partem anatomicam isagoge*, which was published only after his death, in 1555, the Parisian anatomist Jacobus Sylvius (Jacques Dubois, 1478–1555) did offer a fairly precise description of venous valves. He probably was familiar with Estienne's observation: he also mentioned venous valves of the liver and used the same term "epiphysis", which Estienne had used in this context. Sylvius also described a "membrana epiphysis" in the orifice ("in ore") of the azygos vein, however, and he reported having found it also in other places, at the beginning of the great jugular, the brachial, and the leg veins, as well as in the large branch of the *vena cava* originating in the liver. He compared these structures to the "membranes" that closed the orifices of the heart.[188] When exactly Sylvius first saw these valves and possibly showed them to his students is not known. It may have been before Canani's discovery or only afterward, possibly knowing of Canani's observations. In any case, it remains to be noted that Sylvius, who went down in history primarily as a conservative follower of Galen and was heavily criticized by his student Vesalius, was the first to find the venous valves in the large veins and in the *vena azygos*. And Fabrizi clearly was not the first to identify and describe the venous valves. Whether he rediscovered them on his own or – which is quite likely – learned or heard about them from others, for example, from Vesalius' *Examen*, is open to debate.

Muscles

The muscles held a special place in the work of Renaissance anatomists. Vesalius devoted a major part of his *Fabrica* to them. At first sight, this may come as a surprise. Most muscles were only important for the moving of the fingers, hands, arms, and other parts. Their study did not contribute to a better understanding of the physiological processes or the genesis of diseases inside the body. The reason why they nevertheless attracted particular attention was that they are not only very numerous but could also be more conveniently studied than many other parts, except for the bones, because they do not decay and putrefy as quickly.

Falloppia made some important contributions of his own to myology. We have already discussed his findings about the eye muscles, his discovery of the *m. levator palpebri* and his disproval of a *m. retractor bulbi* in humans. He also offered a detailed account and new findings on the muscles of the ear, the jaw, the larynx, and the pharynx.[189]

One muscle, which Falloppia is believed to have discovered and described, for the first time, deserves a special mentioning. The *musculus pyramidalis* not only carries the name Falloppia gave it but to this day is also known as the "Fallopian muscle". It is a small slightly oblique muscle that extends on both sides from the pubic bone to the *linea alba*, which, located between the straight abdominal muscles, connects the sternum with the pubic bone. Falloppia published his discovery in the *Observationes*, arguing at length that it was indeed a separate muscle and not just a part or extension of the straight abdominal muscle.[190] In the early eighteenth century, Giovanni Maria Lancisi raised some doubts whether Falloppia could truly be considered the discoverer. He had found the muscle on one of the (surviving) anatomical tables of Bartolommeo Eustachio, which Lancisi published in print, for the first time, in 1714.[191] An entry in a collection of anatomical notes in the Biblioteca comunale in Siena, which according to the title page were taken by Eustachio, point into the same direction. The page devoted to the abdominal muscles quotes Galen but also refers to a pair of small muscles towards the end of the *musculi recti*, close to the pubic bone, which no one had noticed before.[192] These notes carry no date but the *Tabulae* are believed to have been made in 1552 and Falloppia could well have heard of the muscle from students or scholars who saw it with Eustachio. This would seem untypical, however. Falloppia did not hesitate to attribute the discovery of the stapes to Ingrassia, even though Ingrassia had not published his finding. Handsch's manuscript notes throw new light on the issue. In a section of his notebook on the muscles "according to the demonstration of Gabriele Fallopius" ("secundum demonstrationem Gabrielis Fallopij"), he offers the earliest known written description of this muscle and discussed its uses; according to Falloppia, the muscles supported the excretion of urine.[193] This was in 1552 or, at the latest in 1553 (when Handsch left Padua). His notes leave little doubt that Falloppia found the *musculi pyramidales* either before Eustachio, quite possibly in Pisa already, or at least independently from him. Of course, Eustachio – we only have is anatomical illustrations – could also have heard of Falloppia's discovery in turn.

Vesalius' response

Falloppia's *Observationes* was conceived as an epistolary treatise written to Pietro Manna but the main addressee and target was Vesalius. In general, Falloppia saw himself more on the side of Vesalius than on that of Galen but he also corrected Vesalius in numerous places. Compared to the hostile, aggressive language of some contemporary physicians and naturalists, Falloppia's tone was decidedly moderate. Again and again, he praised Vesalius as "divine". The greatest slight Falloppia inflicted on Vesalius was still subtly clothed in seemingly laudatory words: he praised Vesalius as the one who had perfected the art of anatomy, which Berengario da Carpi, as the "primus restaurator" had restored. Berengario, not the revered Vesalius, it is implied, was the one who had brought back anatomy to its former greatness.[194]

If Falloppia did not seek the open confrontation with Vesalius, this may have been an expression of an overall irenic temperament. He probably was also aware, however, that a sharp and possibly arrogant attack on the famous Vesalius might backfire and damage his own reputation. The student notes on Falloppia's lectures and anatomical demonstrations indicate, in fact, that he was frequently much more outspoken in his oral teaching. Repeatedly Handsch highlighted his notes on what he learned from Falloppia with an "error Vesalii" in the margin.[195] He also added a remark to his previous notes on Fracanzano's account of the branching off of the *vena humeraria*, from the *vena iugularis* explaining that this was the description Vesalius had given but it was false ("sed falsa est"). The *vena humeraria* and the *vena axillaris* had a common origin from a branch of the *vena cava*.[196] The Helmstedt Anonymous even illustrated Vesalius' error in a little drawing which opposed Vesalius' description of the "truth"; he probably copied it from a sketch that Falloppia made for them (see Figure 2.1).

The top half showed, "according to truth in most cases" ("secundum veritatem vtplurimum ita se habet"), how the *vena humeraria* or *cephalica* and the *vena axillaris* branched off right next to each other. The bottom half illustrated Vesalius' claim ("secundum Vesalium hoc modo") that the *vena humeraria* branched off from the *vena iugularis externa*. Which was true, he added, for monkeys but not for humans.[197]

In the *Compendium anatomiae* of 1571, which to all appearances was likewise based on student notes, passages are even more frequently highlighted with "Error Vesalii"[198] and Falloppia's comments on Vesalius are rendered with expressions like "Vesalius was deceived when he said [...]".[199] This includes the repeated accusation that Vesalius had relied on the dissection of animals without making this known, for example, the "error Vesalii" in his account of the *musculus cremaster*, where he was "deceived", in describing its anatomy in the dog.[200] Other passages moreover document Falloppia doubting Vesalius' anatomical skills or at least his diligence. For example,

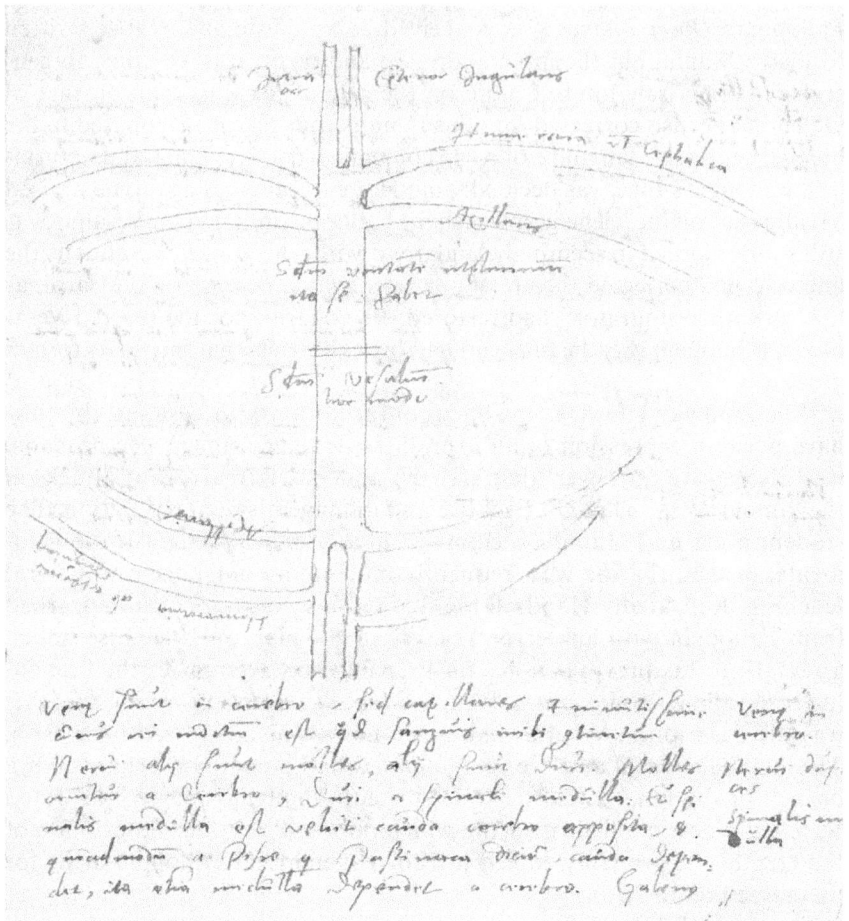

Figure 2.1 Student's drawing showing the branching off of the humeral vein
"according to Vesalius" (below) and "according to truth" (above),
Staats- und Universitätsbibliothek Göttingen, Ms. Meibom 20, fol. 133r.

he emphasized the discovery of a fifth pair of abdominal muscles that Vesalius had overlooked and which would have become obvious with a "more
careful dissection" ("sectione diligentiore").[201]

Vesalius' errors were also a topic of conversation outside the lecture
room. The Helmstedt Anonymus reported in May 1553 that they had been
talking about Vesalius at the home of an acquaintance and that Falloppia
had pointed out to them an "error" Vesalius made concerning the finger
muscles.[202]

Vesalius responded at length to Falloppia's *Observationes*. In 1564, shortly
before Vesalius' death, his *Anatomicarum Gabrielis Falloppii observationum*

examen appeared in Venice.[203] Like Falloppia's *Observationes*, Vesalius' *Examen* has the form of a letter, which he addressed to Falloppia at the court in Madrid on 27 December 1561. With its over 170 pages, it offers a scholarly treatise but it is not certain that Vesalius conceived it for publication. As the printer Franciscus de Franciscis explained in his preface to the reader, Vesalius had stopped in Venice on his way to Jerusalem – a journey from which he was never to return. On that occasion, Vesalius had explained to the printer and to various physicians who happened to be present in his print shop that he had given his response to Falloppia's *Observationes* to the Venetian envoy Paolo Teupolo (= Tiepolo), when he was still in Spain. Tiepolo, however, was delayed for many months in Catalonia because of the war negotiations in France and also because he could not find a galley that would take him to Italy by sea. When he finally arrived in Venice, Falloppia was already dead. Tiepolo therefore kept the manuscript and it could easily be obtained from him. Since various people present requested a copy, it was agreed, according to the printer, that he be given the manuscript for publication.[204] If he had wanted to publish it, Vesalius could easily have put the text into print much earlier, in Spain. It is conceivable, however, that he hoped Falloppia would include his exam in future editions of his *Observationes*: Vesalius expressly suggested to Falloppia that he make his writing available to the readers of his *Observationes*.[205]

In hindsight, Vesalius' *Examen* had very little to offer that was new.[206] He mostly just repeated what he had already written in his *De fabrica*. Vesalius himself expressed his regret that he had no opportunity in Madrid (where he had moved from Brussels in 1559) to dissect cadavers and conduct anatomical research; he could not even easily get hold of a skull. He hoped to find the opportunity for further anatomical research in the future and promised to report his findings to Falloppia immediately.[207]

For historical analysis, Vesalius' response is primarily revealing in terms of how he dealt with Falloppia's challenge. Vesalius could become quite unpleasant with colleagues. His judgment of Juan de Valverde, for example, was scathing. He had never dissected himself and had no idea about medicine, Vesalius claimed.[208] Unlike in the case of Valverde or the already deceased Realdo Colombo, Vesalius may well have recognized the potential dangers to his fame, however, of a personal attack on Falloppia. If Falloppia eventually published the comprehensive and richly illustrated textbook of anatomy he announced in the *Observationes*, he would have no reason to have regard for Vesalius' reputation and could harshly expose his gaps and errors. Just as Falloppia had done to him before, Vesalius chose a decidedly friendly tone instead. Falloppia's *Observationes* were very welcome to him, he explained by way of introduction, since Falloppia was an extremely skilled anatomist. Instead of attacking Falloppia directly, he subtly played down the importance of his work. If Falloppia made the *Examen* available to the readers of his *Observationes*, Vesalius declared, Falloppia would help them acquire a more comprehensive knowledge of anatomy – if not to say an appendix to Vesalius' *Fabrica*.[209]

The well-known anatomist and surgeon Johannes Jessenius took the author of the *Fabrica* at his word. He added the subtitle "Appendix to the great work on the fabric of the human body" to his new edition of Vesalius' *Examen* in 1609. In his dedicatory letter to the Duke of Brunswick and Lüneburg, he also indicated, however, that Falloppia's *Observationes* did have an impact on Vesalius' reputation. Falloppia, Jessenius declared, was of no less sharp a mind than his teacher Vesalius (apparently, like many historians later, he mistook Falloppia for a personal student of Vesalius). He differed on quite a few points from Vesalius and, while emulating the work of Vesalius, the *Observationes* had diminished the esteem in which Vesalius was held.[210]

In the long run, Vesalius' Fabrica was to remain the great iconic work of Renaissance anatomy, especially thanks to its outstanding illustrations. Had Falloppia lived long enough to realize his own project of a comprehensive, illustrated textbook, two great iconic anatomical works would possibly stand side by side today, Vesalius' *Fabrica* and Falloppia's work which, as the first atlas of topographical anatomy, would have offered illustrations that not only showed the various parts of the human body in but also in their spatial relationship to other parts as they could be observed in situ.

Comparative anatomy

Falloppia frequently resorted to animals in his teaching, to show structures like the muscle that raises the upper eyelid or changes like those in the pregnant uterus which could not easily be demonstrated on humans or were harder to see in humans because of the smaller size of the respective part. Falloppia's interest in animal anatomy went further than that, however. He has a major – and so far underestimated – place in the history of comparative anatomy. He was one of the very first anatomists to study the relationship between structure or morphology and function by looking at how the parts of the body were shaped differently in different animals to allow them to fulfill the same function.

In his classic overview of the history of comparative anatomy, Francis J. Cole understandably mentioned Falloppia in passing only.[211] Falloppia's references to animal anatomy in the *Observationes anatomicae* rarely went beyond merely pointing out similarities in fact. The notes Georg Handsch and Helmstedt Anonymus made on Falloppia's anatomical demonstrations in the early 1550s already offer a different picture, however. Falloppia not only pointed out the differences but also sought to explain why the different shape or form of certain parts or organs was appropriate and suitable for that specific animal. Today, his arguments remind us of the explanations of evolutionary biologists, who attribute anatomical differences to corresponding functional advantages for the animal in its respective habitat.

Falloppia's explanations show that one could reach quite similar conclusions without assuming a historical, evolutionary development. His conceptual basis was a profoundly teleological understanding of nature and man, which Renaissance physicians inherited and adapted from Aristotle and Galen. In the divine creation or, in secular terms, in nature, everything had its meaning and purpose. The structures that anatomists found in humans and animals thus always had a certain purpose, a certain function, which the anatomist had to determine. If certain anatomical structures were shaped differently or could even only be found in some animals and were missing in others, this had to be explained by their function, by the tasks for which they were needed – or not.

Perhaps the most famous fallacy of Galen's anatomy which arose from the fact that he based his account on the dissection of animals was his account of a *rete mirabile* at the base of the human brain. This *rete mirabile*, a conglomerate of arterial and venous vessels, is found in various vertebrate animals but not in humans. Falloppia not only explained to his students that the *rete mirabile* or *plexus retiformis*, as he also called it, did not occur in humans but was an "exquisite" part in horses and cattle. He also explained why they had a *rete* mirabile and man did not. When horses were exercising and running or cattle pulling carts or ploughs, they got hot and much venous and arterial blood ascended toward the head where it usually got mixed with the air inhaled through the nose to make animal spirit. When the arterial blood arrived in too large a quantity and very suddenly at the head, however, it could suffocate the animal spirit. Therefore, nature made this *plexus reticularis*, which could accommodate a lot of blood in its numerous and convoluted vessels and softened the impact of the blood on the cerebral ventricles.[212]

Another example of Falloppia's comparative approach is his explanation why the *musculus retractor bulbi* could be found in cattle but not in man. Unlike humans with their upright position, quadrupeds had their eyes directed more downward, toward the ground. If their eyes were not anchored firmly, they threatened to prolapse. The students could indirectly experience this in their own body. If one kept the head lowered for a longer period of time, one could feel a certain heaviness in the eyes, as if they were about to prolapse, because just that muscle was missing.[213]

The comparison between humans, who walked upright, and other animals also made Falloppia reject the traditional explanation of the function or use of the pancreas. The anatomists, he explained, claimed that this *glandula* served as a support for the stomach above it. But in that case the pancreas would be useless in animals. They moved on four legs, which put the pancreas above the stomach not underneath it. The true use of the pancreas was a different one: it provided a safe pathway for the big vein that ran from the liver to the spleen.[214]

The existence of a tongue in many animals, similar in the substance of its muscles and fibers to that of humans, raised the question, in turn, why only humans could speak. One reason, Falloppia argued, was that animals lacked the intellect that was necessary for imitation. But it was also a matter of size and shape, Falloppia explained. If the tongue was long, as in dogs, or pointed, as in birds, or thick, as in cattle, it was not suitable for articulation.[215]

In his demonstration of the caecum of a monkey with the valvular structure between the small intestines and the colon he had discovered, Falloppia even took recourse to arguments from embryological development. Cats, mice, pigs, and other animals, he explained, walked on four feet and sometimes even down walls. The contents of the colon would therefore easily regurgitate into the small intestines if they were not held back.[216] In monkeys, he claimed, the part was even larger than in pigs because they jumped down from the trees and frequently vehemently lunged downward.[217] Apparently, he thought that the larger the caecum and the fecal mass it could accommodate, the better it could serve as a reservoir and ease the pressure on the valvular structure. In humans, by contrast it was of little or no importance because they walked upright. His audience might ask why nature made this part in humans, since nature did nothing in vain. The answer was that, just like the umbilical vessels no longer were of use in grownups but necessary for the nutrition of the fetus, the caecum – and the valvular structure it is implied – was very useful in the fetus who came out of the womb head first. For this reason, this part was relatively much bigger in the newborn than in grownups.[218]

Falloppia's explanations may not convince us today but this is not the point. What is decisive in our context is his methodological approach. He did not just describe similarities of the parts of the body between humans and animals. He also sought to explain the differences by looking at the function these parts fulfilled – or could not or did not need to fulfill – in animals with their specific build and in their respective habitat.

Notes

1 See, in particular, Carlino, Books (1999); Mandressi, Le regard (2003).
2 Cunningham, Anatomical Renaissance (1997).
3 Klestinec, Theaters (2011); Stolberg, Teaching anatomy (2018); idem, Learning anatomy (2018).
4 Carlino, Books (1999).
5 ÖNBW, Cod. 11210, with Georg Handsch's notes on an anatomical demonstration.
6 See also Heseler's notes on the anatomical demonstration bei Corti and Vesalius in Padua in 1540 (Eriksson, Vesalius' first public anatomy (1959)).
7 ÖNBW, Cod. 11210, fol. 27r.
8 SUBG, Ms. Meibom 20, foll. 144r–240r.
9 Ibid., foll. 83r–126v, *De partibus similaribus*, begun on 4 December 1552; the notes end rather abruptly with the heading *De arteria* and are probably incomplete.
10 Ibid., foll. 1r–37r and ibid., foll. 38r–71v, notes Falloppia's lectures on *De ossibus* in 1551 and 1552; Georg Handsch attended the same lecture in 1551 (ÖNBW, Cod. 11210, fol. 34v: "Anno 1551 Anatomicus Gabriel legit libellum Galeni de ossibus".

11 SUBG, Ms. Meibom 20, foll. 83r–126v, *De partibus similaribus*, begun on 4 December 1552; the notes end rather abruptly with the heading *De arteria* and thus are probably incomplete; Georg Handsch added a piece of paper with brief notes on *De membris similaribus* to his notes on the anatomical demonstration by Antonio Fracanzano (ÖNBW, Cod. 11210, fol. 188r–v); he does not indicate his source but it seems very likely that these notes likewise reflect a lecture by Falloppia on *De partibus similaribus* – and quite possible the same that was documented by the Helmstedt Anonymous. This conclusion is supported by the fact that the lecturer stressed, as Falloppia did in his anatomical lectures, that no nerv – and against Galen's claim not even the optical nerv – was hollow.

12 Falloppia, De partibus (1577), *proemium* by Seidel; Seidel claimed that Falloppia lectured several times ("multoties") on this topic. So there may even have been more than three lectures.

13 SUBG, Ms. Meibom 20, fol. 36r: "Haec lectio a paucis audita fuit".

14 For a detailed contemporary account see Colombo, Re anatomica (1559), pp. 110–117.

15 Cit. in Favaro, Gabrielle Falloppia (1928), pp. 212–213.

16 Kornell, Illustrations (2000); Canalis, Vesalius' methods (2018), p. 138.

17 ÖNBW, Cod. 11210, fol. 191v.

18 Ibid., foll. 30v–31v.

19 Ibid., fol. 31r: "Monstravit etiam cranium canis et simiae."

20 Vesalius, De humani corporis (1543), p. 547; cf. Carlino, Books (1999), pp. 188–189.

21 ÖNBW, Cod. 11210, fol. 193r.

22 Ibid., Cod. 11210, fol. 192v.

23 Ibid.

24 Ibid., fol. 194r.

25 Ibid., fol. 192v. In her work on the teaching of anatomy in sixteenth-century Padua, Cynthia Klestinec described the interest in practical anatomical skills as a new development in the late sixteenth century (Klestinec, Theaters (2011), pp. 153–155 and p. 225, note). Handsch's student notes provide a much earlier evidence for this trend, which is not well documented in Klestinec's principal source, the *Acta* of the *Natio germanica*.

26 Stolberg, Training (2022).

27 There must have been at least two corpses since he concluded by asking his students to pray to God to have mercy on "their soul" ("istorum animum") (SUBG, Ms. Meibom 20, fol. 236v).

28 SUBG, Ms. Meibom 20, foll. 236v–240r.

29 According to Falloppia, Observationes (1561), fol. 64r.

30 SUBG, Ms. Meibom 20, foll. 136r.

31 ÖNBW, Cod. 11210, fol. 193v.

32 SUBG, Ms. Meibom 20, foll. 138v–143v, continued on foll. 127r–135r.

33 Falloppia, Observationes anathomicae (1570), fol. 74r, "in uno, ac altero cadauere eo anno 1554".

34 SUBG, Ms. Meibom 20, fol. 248r.

35 See the chapter on conflicts and tensions.

36 Schieß, Briefe (1906), p. 20.

37 Biblioteca Universitaria di Bologna, Mss Aldrovandi, 38², I, fol. 43, letter from Falloppia to Ulisse Aldrovandi, Padua, 29 May 1556, ed. in Di Pietro, Epistolario (1970), pp. 27–28.

38 Letter from Falloppia to the *Riformatori*, Padua, 12 December 1556 in ASV, Lettere dei Riformatori allo Studio, 1555–1559, filza 63; ed. in Di Pietro, Epistolario (1970), pp. 29–30.

39 Letter from Falloppia to Aldrovandi, 30 January 1559, ed. in Di Pietro, Epistolario (1970), pp. 31–32.

40 Ferinarius, Narratio (1601), no pagination.
41 Letter from Falloppia to Aldrovandi, 30 January 1559, ed. in Di Pietro, Epistolario (1970), pp. 47–48.
42 Favaro, Atti (1911), p. 29: "Verum nihil effectum est partim detrectante operam Fallopio anatomico et valetudinem excusante, partim ipso Rectore connivente et non urgente."
43 Ibid., pp. 34–35, "nisi et tempus aliquantisper contrarium nobis fuisset et in medio quasi deficiens minime firma uteretur valetudine, quo nimirum factum est ut postea ad finem nimis fuerit properatum."
44 Ibid., p. 41: "suam valetudinem, quam praetendebat, potius curans quam utilitatem studiosorum respiciens, quamvis saepius rogatus et praesertim a nostris, administrare renueret, tum quia Reformatoribus persuassisset, sufficere si tertio saltem anno anatome administraretur."
45 See my detailed account of the concepts that guided everyday medical practice at the time in Stolberg, Learned physicians (2022).
46 ÖNBW, Cod. 11210, foll. 56v–57r.
47 Falloppia, De vulneribus (1606), p. 236.
48 Ibid., p. 399, "locus quoque, in quo est vulnus, ducet vos in cognitionem an vulneratus sit neruus aliquis: et hic oportet vos esse exercitatos in anatome: quia ita sciueritis, per quam partem percurrant nerui, et per quam non."
49 ÖNBW, Cod. 11210, fol. 14r.
50 Falloppia, Compendium (1585), foll. 42v–43r.
51 ÖNBW, Cod. 11210, fol. 207v.
52 SUBG, Ms. Meibom 20, fol. 140r; ÖNBW, Cod. 11210, fol. 199v.
53 ÖNBW, Cod. 11210, fol. 4v.
54 SUBG, Ms. Meibom 20, fol. 129r; ÖNBW, Cod. 11210, fol. 14r.
55 Falloppia, De ossibus (1570), foll. 75v–76r.
56 Ibid., fol. 75v.
57 Stolberg, Bedside teaching (2014).
58 Archivio di Stato di Firenze, fondo Guidi 571, letter from Falloppia to Francesco Torelli, Pisa, 5 January 1550.
59 ÖNBW, Cod. 11240, fol. 78r.
60 Archivio di Stato di Firenze, fondo Guidi 571, letter from Falloppia to Giacopo Guidi, Pisa, 6 November 1550, complaining about the situation in the previous winter.
61 Archivio di Stato di Padova, Archivio civico antico, Ducali 6, fol. 42r. The letter is dated 11 February 1549 but this was apparently according to Venetian style (with the new year starting in March only) because it carries an administrative note that is was received on 20 February 1550.
62 Presumably Falloppia meant that the last public anatomy had taken place two years before, in the winter of 1554/55, which is well documented due to protests of the students against Trincavella's lecturing.
63 Letter from Falloppia to the *Riformatori*, Padua, 12 December 1556 in ASV, Lettere dei Riformatori allo Studio, 1555–1559, filza 63, ed. in Favaro, Gabrielle Falloppia (1928), pp. 226–227 and Di Pietro, Epistolario (1970), pp. 29–30.
64 Draft of a letter from the *Riformatori* to the *Podestà* in Padua, 15 December 1556, in ASV, Riformatori allo studio, filza 63, ed. in Favaro, Gabrielle Falloppia (1928), p. 227.
65 Letter from Falloppia to Lodovico Corbinelli in Ancona, Padua, 29 January 1557, first published by Angelini, Una lettera (1900), who found the original in the Archivio di Stato di Firenze and also mentioned four other letters from Falloppia to Corbinelli but did not provide a proper reference. Di Pietro could

not find the originals and reprinted Angelini's edition (Di Pietro, Epistolario (1970), pp. 30–32); neither have I been able to identify the originals so far. The Archive's holdings on the "Patrimonio ecclesiastico" refer to the late eighteenth century today.

66 Drafts of letters from the *Riformatori* Padua, 7 and 15 December 1557 in ASV, Riformatori allo studio, filza 63.

67 Hyrtl, Lehrbuch (1846), p. 22.

68 Andreozzi, Leggi (1878) pp. 43–45; Ambrose, Immunology's first priority dispute (2006), p. 3; Öncel, One of the great pioneers (2016).

69 Conde Parrado, Entre la ambigüedad y la audacia (1999), p. 18.

70 O'Malley, Gabriele Falloppia's account (1968); idem, Falloppio (2008), p. 519.

71 Falloppia, Libelli (1563), foll. 47v–48r. The printer Donato Bertelli explained in his dedicatory letter that the two treatises had come into his hands and were found to be as pure and simple as if the author had compiled and written them with his own hand ("in manus nostras devenerunt, ita pura quidem, atque simplicia ut ab autore ipso compilata, manuque propria conscripta, fuerunt inventa").

72 Falloppia, Libelli (1563), fol. 48r:

> Nam Princeps iubet ut nobis dent hominem, quem nostro modo interficimus, et illum anatomizamus cui exhibui 3 ii. opij. & adueniens paroxismus (nam hic patiebatur quartanam) prohibuit opij actionem. Hic gloriabundus rogauit, ut bis adhuc exhiberemus, quod si non moreretur, ut procuraremus pro eius salute apud Principem. rursus illi exhibuimus extra paroxysmum 3 ii. opij, et mortuus est.

The same passage was published verbatim in Falloppia, Opera (1584), p. 712, except that the printer used "drach[ma]" instead of the sign 3.

73 Falloppia, De compositione (1570); the dedicatory letter by the printer Paulus Meietus does not indicate from whom he received these notes.

74 Ibid., fol. 13r: "dicam quid accidit, cum essem Pisis, Dux n[oster] corpora iustitiae tradenda, anatomicis exhibebat, ut morte, quam sibi videbatur, ab ipsis interficerentur: nos autem uni exhibuimus drachmam opii, et spatium septem horarum ipsum interfecimus, exhibuimus autem quandoque alteri laboranti quartana, qui post potum statim rigore affici coepit, cui maximus calor inde successit, neque mortuus est, utpote victo a calore naturale ipso opio"; the same passage can be found verbatim in the various editions of Falloppia's works (Falloppia, Opera (1584), p. 157; Falloppia, Opera (1600), vol. 1, p. 141; Falloppia, Opera (1606), vol. 3, fol. 200v) except that the "n." after "Dux" – which almost certainly stood for "noster" ("our") – is rendered as "enim". The passage can still be found, in 1675, in an Italian edition of Falloppia's surgery (Falloppia, Chirurgia (Venice: Curti 1675), p. 36).

75 Astruc, De morbis (1740), pp. 748f: "barbaram illam crudelitatem Falloppii, medici christiani, medici seculo decimo sexto nati, qui carnificis partes aperte sustinere non exhorruerit."

76 Roth, Andreas Vesalius (1892), pp. 477–485.

77 Andreozzi, Leggi (1878), p. 53: "la cosegni secondo il solito al nototomista per farne anatomia."

78 Anonymous, Memorie (1756), p. 39; Fabroni, Historia (1792), p. 78.

79 ÖNBW, Cod. 11225, fol. 36r: "Nam ego vidi cuidam abiudicati datas 3 ii opii in paroxismi febris quartanae, et febris vicit venenum et liberatus est, secundo die iterum datum opium mortuus est sicut etiam alii." This is one of the student notebooks of Georg Handsch.

80 ÖNBW, Cod. 11240, fol. 78r:

Falloppius narravit de novem abiudicatis quibus dedit opium, unicuilibet dr. ii in vino malvatico et adiuncto aliquanto Diacodii pro dulcoratione. Insumpserunt et paulopost dormiverunt ad 4 horas, deinde petierunt potum et datus est eis, et rursus dormiverunt per 8 horas et sic dormiendo exhalarunt. Verum omnes sudarunt nec insecuta sunt ulla symptomata qualia recenset Dioscorides. Unus fuit quartanarius et ea nocte cum debebat habere paroxysmum eo vesperi exhibuit ei, et non mortuus est: c[ausa?] opto forsan quia calor febrilis obstitit. Verum cum alio die exhiberet secundo etiam obiit. Dictum est de dr[achmis] ii, non tamen totum ebiberunt, quia maior pars residebat in fundo.

81 ÖNBW, Cod. 11205, fol. 223r: "Nota hystoriam Fallopii, qui dedit 9. abiudicatis opium, et unus ex illis quartanarius non mortuus est."
82 ÖNBW, Cod. 11207, fol. 94v.
83 On 17 September 1551, it was ordered that a certain Maria Maddalena di Pieraccioli from Modigliana was to remain in the Bargello. Eventually she was taken to Pisa but this may well have been only after Falloppia's departure for Padua.
84 Falloppia, Chirurgia (1603), fol. 31v.
85 Stolberg, Tödliche Menschenversuche (2014); Rankin, Poison trials (2021).
86 Juan Fragoso, *Cirugia universal*. 6th edn. Alcalá de Henares: Juan de Gradan 1608, p. 159, quoted in Conde Parredo, Entre la ambigüedad y la audacia (1999), p. 20.
87 *Observationes anatomicae ad Petrum Mannam*. Venice: apud Marcum Antonium Ulmum 1561.
88 Falloppia, Observationes (1561), letter to the reader: "Librum hunc ante quattuor annos scripseram."
89 Cologne: apud haeredes A. Birckmanni 1562; Paris: apud Iacobum Kerver 1562.
90 *Observationes anatomicae in quinque libros digestae*. Ed. by Johannes Siegfried. Helmstedt: Lucius 1588.
91 Colombo, Re anatomica (1559), p. 257.
92 Koch, Schüler Vesals (1972), p. 67.
93 Sacchi, Intorno agli Anguissola (1994), pp. 354–356. Today, the painting is in the Museo del Prado in Madrid.
94 Rossi, Medaglie (1985), p. 351.
95 Sacchi, Intorno agli Anguissola (1994), esp. pp. 354–356; Sacchi bases her conclusions convincingly on two documents in the Archivio di Stato in Milan, namely Autografi 216, fasc. 34, "Manna Pietro" and Sanità p.a. 209.
96 Haller, Bibliotheca anatomica (1774), p. 218.
97 Falloppia, Observationes (1561), "ad lectorem".
98 As Muratori and Bighi, Andrea Vesalio (1963–1964), pp. 72–73, stated bluntly: "Vesalio ripete assai spesso gli errori di Galeno di estendere alla specie umana le disposizioni anatomiche che si riscontrano nei mammiferi."
99 Falloppia, Observationes (1561), "ad lectorem".
100 Colombo, De re anatomica (1559), p. 243; see also the chapter on the female genitals.
101 Falloppia's student Andrea Marcolino also mentioned Falloppia's plan (dedicatory epistle to Cardinal Aloysius d'Este in Falloppia, *De simplicibus medicamentis purgantibus* (1555), reprinted also in the 1556 edition).
102 Falloppia, Observationes (1561), fol. 6r-v:

praecipue cum volumen anatomicum moliar, in quo anatomen integram complectar, atque minutissima quaeque ad hanc artem pertinentia. Illud erit quamplurimis figuris exornatum, quae corpus humanum, et simias etiam, vt Galeni dogmata faciliora sint, exprimant. Neque hae imagines solum eiusdem generis partes, ac similes inter se continebunt, sed etiam diuersis ex classibus inter se simul complicatas, veluti in ipsomet corpore humano connectuntur, indicabunt.

103 Letter from Alfonso II to Renato Brasavola, 12 October 1562, ed. in Tiraboschi, Biblioteca modenese (1782), pp. 245–246.
104 Universitätsbibliothek Erlangen, Trew, Purkircher, Nr. 2, letter from Georg Purkircher to Joachim Camerarius, Padua, 26 November 1562.
105 Wieland eventually bequeathed his books but not his manuscripts to the City of Venice and they can today be found in the Biblioteca Marciana (Ferrari, Opere (1959)).
106 Falloppia, Observationes anathomicae (1570); in the various editions of Falloppia's *Opera omnia* they were then published somewhat misleadingly as "Observationes de venis".
107 Falloppia, *De humani corporis anatome compendium* (1571); ibid., fol. 4r, gives the title "anatomicae institutiones".
108 Falloppia, *De humani corporis anatome compendium* (1585).
109 A special case is Achille Olivieri (Olivieri, Experimentum (2006)), who, totally ignoring the *Observationes*, sought to reconstruct Falloppia's epistemology and experimental terminology, drawing among others on the work of Michel Foucault, just from the *Compendium*, undoubtedly based on the wrong assumption that it was written (and, as he claimed, dedicated to Fabrizi d'Acquapendente) by Falloppia himself.
110 Haller, Bibliotheca anatomica (1774), p. 220.
111 Ibid.
112 An example is ibid., pp. 218–221; the first detailed and extensive account was given by Calderato, Brevi cenni (1862), pp. 19–48.
113 Falloppia, Observationes (1561), foll. 39r–42v; cf. Gysel, Gabriele Falloppio (1973).
114 Herrlinger/Feiner, Why did Vesalius not discover the Fallopian tubes? (1964).
115 ÖNBW, Cod. 11210, foll. 11r–12v; Handsch's notes begin on fol. 11r, with a "Gabriel Falloppius" added to the heading "De conceptu". It is not quite clear, from Handsch's notes, where this part of his notes ends but by all appearances also the following section on the fetal vessels and the fetal membranes and on giving birth record Falloppia's teaching and not that of Fracanzano into which these notes were inserted. The notes on the pages from fol. 13r carry the neatly written heading "Secunda pars anatomiae de thorace sive pectore" and clearly document Fracanzano's teaching again.
116 SUBG, Ms. Meibom 20, fol. 237r: "Haec sunt ridicula". In his book *Making sex*, Thomas Laqueur claimed that learned medicine down to the eighteenth century adhered to a "one-sex model" of the human body and described the male and female genitals as merely differing in their location. As various historians, including myself, have shown, Laqueur's highly influential claim is based on his almost complete ignorance of the predominantly Latin medical and anatomical literature of the sixteenth and seventeenth centuries and does in no way justice to the range of views on anatomical sex difference that can be found already in ancient and medieval writing; cf. Park and Robert, Destiny is anatomy (1991); Stolberg, A woman down to her bones (2003); King, The one-sex body on trial (2013).
117 SUBG, Ms. Meibom 20, fol. 237r–v, "Ad hos testes descendunt arteriae et venae ut in viris quae deferunt sanguinem: habent et processus varicosos breves continentes semen, ex quibus emittunt in fundum uteri."
118 Falloppia, Compendium (1571), fol. 23r–v:

> talis est hactenus decantata ab omnibus anatomicis vteri vasorum descriptio, quam dum accuratius examinare tentamus, inuenimus vasa illa, quae semen

ferre opinati sunt ipsi, nullam prorsus cum testibus communicantiam habere, praeter fibrosa et membranea quaedam ligamenta, quae vasorum illorum principium testibus et notabili satis interstitio iungunt: vasa ista seminalium vasorum imaginem exprimentia, qualis in masculis reperitur, ab vtero nascuntur, sursumque oblique ascendentia e [sic!] regione testium, sed exterius magis in cauitatem quandam desinunt latiorem ibi, quam ipsorum vasorum cauitas sit. in capite enim dilatatur membrana, ita ut quandam vesiculam constituat, quae semper rugosa admodum reperitur, et variculae cuiusdam speciem referens, qualis est illorum vasorum, quae in viris ad vesicae collum semen perfectum continent. at cauitas haec osculum habet, quod apertum semper reperitur, ita vt ab eo specillum in vteri cauitatem peruenire possit, nulla interim in parte laesis aut laceratis vasis ipsis: vasa ista ita se habent, ac si duorum spiraculorum aut caminorum vicem gererent, per quos vteri fuligines intra abdominis cauitatem exhalarent. haec an praecipua sit istorum vasorum utilitas, non satis affirmare ausim, cum demum per frequentiorem diligentioremque sectionem ipsorum naturam edocti erimus, accuratius de ipsis pleniusqe afferre enitemur.

119 Falloppia, Observationes (1561), fol. 195v:

Omnes anatomici uno ore asserunt in testibus foeminarum semen fieri, et quod semine referti reperiantur, quod ego nunquam uidere potui, quanuis [sic!] non leuem operam ut hoc cognoscerem adhibuerim. Vidi quidem in ipsis quasdam ueluti uesicas aqua uel humore aqueo, alias luteo alias uero lympido turgentes. Sed nunquam semen uidi, nisi in uasis ipsis spermaticis uel delatoriis uocatis.

120 Ibid., foll. 195v–196r:

Aliud asserunt, quod uasa ista spermatica oriuntur a testibus atque cum illis omnino copulantur, et desinunt in cornua ipsius vteri ita appellata; quod minime placet, quoniam nunquam observare potui meatus istos seminarios coniunctos cum testibus, nisi uterus male affectus fuerit; in omnibus autem aliis, aut virgines fuerint, aut uitiatae mulieres, uel ediderint conceptum foetum aut non, semper disiunctos uidi testes a dictis meatibus, neque ulla apparet uena, aut angustum uas, uel magnum, quod a testibus ad hos meatus transeat, solaque adest membranula tenuissima a peritonaeo, quae cum omnia ibi contenta uasa uenas scilicet et arterias uinciat, hos quoque meatus et testes simul copulat, ita tamen ut non se inuicem tangant. Nam inter ipsos spacium ferme dimidii digiti transuersi intercedit, tantumque abest, ut simul connascantur.

121 Falloppia, Observationes (1561) foll. 196v–197v:

Meatus uero iste seminarius gracilis et angustus admodum oritur nerueus ac candidus a cornu ipsius uteri, cumque parum recesserit ab eo latior sensim redditur, et capreoli modo crispat se, donec ueniat prope finem, tunc dimissis capreolaribus rugis, atque ualde latus reditus [sic!] finit in extremum quodam, quod membranosum carneumque ob colorem rubrum uidetur, extremumque lacerum ualde et attritum est ueluti sunt pannorum attritorum fimbriae, et foramen amplum habet, quod semper clausum iacet concidentibus fimbriis illis extremis, quae tamen si diligenter aperiantur, ac dilatentur tubae cuiusdam aeneae extremum orificium exprimunt. Quare cum huius classici organi demptis capreolis, uel etiam ijsdem additis meatus seminarius a principio usque ad extremum speciem gerat, ideo a me uteri tuba uocatus est. Ita se haec habent in omnibus, non solum humanis, sed etiam ouinis, ac uacinis cadaueribus, reliquisque brutorum omnium, quae ego secui.

122 The *Compendium* more specifically mentions Vesalius as the one who made this mistake. The passage is highlighted with "error Vesalii" in the margin (Falloppia, Compendium (1571), fol. 22r).

123 Falloppia, Observationes (1561), fol. 192r–v.

124 ÖNBW, Cod. 11210, fol. 11r–v; SUBG, Ms. Meibom 20, fol. 237v.

125 ÖNBW, Cod. 11210, fol. 11r–v; similarly but without mentioning menstrual bleeding in early pregnancy: SUBG, Ms. Meibom 20, fol. 328r.

126 Cf. Berengario da Carpi, Anatomia (1535), fol. 20r: "panniculus virginalis ut rete: contextum ligamentis subtilibus et pluribis venis: quo violata caret, quia rumpit in primo coitu cum mare: hic panniculus vocatur eugion et cento et imen."

127 SUBG, Ms. Meibom 20, foll. 237v–238r; ÖNBW, Cod. 11210, fol. 10r, inserted leaf. Handsch also noted Fracanzano's objection that the hymen could not be found normally but existed as a preternatural condition in some women who for this reason could not menstruate and conceive. In his *Observationes* (1561), fol. 194r, Falloppia attributed this explanation to Soranos.

128 Falloppia, Observationes (1561), foll. 193v–194v.

129 Colombo, De re anatomica (1559), p. 243; Colombo named this part, which, he claimed, noone had ever mentioned, "amor Veneris" or "dulcedo Veneris".

130 Park, Rediscovery (1997), p. 177; Bartholin, Bartholinus anatomy (1668), p. 75.

131 Falloppia, Observationes (1561), fol. 193r–v.

132 Ibid., foll. 192v–193r.

133 As Katherine Park (Park, Rediscovery (1997), p. 176, has pointed out, the first among the Renaissance anatomists who mentioned the clitoris (but connected it with urination) was probably Charles Estienne.

134 Falloppia, Compendium (1571), fol. 22r:

> rima in superiori parte uersus pubem magis corpus quoddam eminens habet, quod in quibusdam mulieribus adeo longum est, ut pudendi cuiusdam uirilis prae se ferat imaginem, praesertim cum uestibus attritum & confricatum turget, atque inflatur virilis pudendi modo: corpus istud eandem conformationis rationem obtinet, quam et virile pudendum.

135 See the chapter on cosmetic medicine below.

136 See also Park, Rediscovery (1977).

137 ÖNBW, Cod. 11210, fol. 197v; Handsch may even have written these notes in 1551, during the first winter he studied with Falloppia. In 1553, when Falloppia demonstrated the *caecum intestinum* on a monkey, Handsch made reference to this previous demonstration in a note that he added in the margin, repeating some of the things he had already written before.

138 Ibid.:

> De coeco intestino. Hunc usum dixit Gabriel, quod [?] cum intestini coli inicium sit crassisimum hoc crassum inicium non directe continuatur ileo, sed dextrorsum declinat in latus, ideo expertus est, cum aquam fundit per gracilia intestina, tum bene descendit per omnia intestina, cum autem fundit per rectum, tum aqua non transfluit omnia intestina, sed sistitur et continetur in coli principio illo crasso. ideo ex hoc eciam [sic] appensum est coecum ut si hoc impleatur et crassi principium, eciam tum ex repletione clauditur quodammodo ileon, et ex hoc usum eius invenit, nempe ne feces regurgitent ad superiora (sicut eciam natura procedit in meatibus urinariis ad vesicam, quae ita continet urinam, ut colon feces, ne urina regurgitet).

139 SUBG, Ms. Meibom 20, fol. 140r:

> Coecum intestinum est finis coli, et ad latus ileon inseritur. [...] Coeci usus est in simiis ne regurgitet cibus ad partes superiores cum prona incedunt:

quodque haec usus sit, signum est: quia si in rectum aqua immittatur, aut flatus perveniat in coecum, non transgreditur a crassa: at si superius immittatur pertransit. ratio est: quia ad insertionem ilei plicae sunt duae, quae in inflatione et repletione comprimuntur ut in corde fit, et prohibent regressum.

140 ÖNBW, Cod. 11210, fol. 197v, note added in the margin.
141 SUBG, Ms. Meibom 20, fol. 140r: "unde nec clysteria possunt pervenire ad partes illas et pertransire, ita ut eijciantur per vomitum in homine, nisi debilibus ex morbo existentibus intestinis"; ÖNBW, Cod. 11210, fol. 197v, note added in the margins.
142 Ambrose, Immunology's first priority dispute (2006); Natale, Boci and Ribatti, Scholars (2017).
143 Gaspare Aselli, De lactibus (1627).
144 ÖNBW, Cod. 11210, fol. 207v:

> Hoc anno in privata anatomia dixit. Esse quosdam meatus prope venam portae, ubi nervus ab orificio ventriculi inseritur qui quendam humorem continent, quem et usum eius se nondum potuisse cognoscere dixit. Monstravit in cane et dixit esse in homine eciam. Nec Galenus nec Vesalius mentionem illius facit et cuidam scripturienti ita dictavit Fallopius. Oriuntur a sima parte hepatis, a latere venae portae, et desinunt in inferiorem tunicam omenti atque se inserunt in pancreas. luteum continent succum nullo sapore exquisito praeditum, substantia tenuem et si aliquo sapore est praeditum, magis in amaritudinem vergit, quam ad alium, quamvis nec illa amaritudo sit exquisita, flavii sunt plurimi, et forte ad pancreas veniunt, ut illud nutriant.

145 Falloppia, Expositio (1570).
146 Ibid., fol. 72v: "In infima parte hepatis sunt quidam parui meatus, qui desinunt, ac terminantur im [sic!] pancreas [sic!] et in glandulas ibi proximas, qui quidem minimi meatus deferunt quendam succum oleaginosum flavum, et tendentem ad aliquam amaritudinem."
147 Ambrose, Immunology's first priority dispute (2006).
148 Berengario da Carpi, Anatomia (1535); cf. Parent, Berengario da Carpi (2019).
149 O'Malley, Gabriele Falloppia's account of the cranial nerves (1961), p. 132.
150 Ibid.; Falloppia, Observationes (1561), foll. 136v–156r.
151 Falloppia, Observationes (1561), fol. 140r–v.
152 Ibid., foll. 147v–148r.
153 Ibid., foll. 148r–151v; cf. Canalis, Gabrielle Falloppia (2018), pp. 187–189 and ibid., pp. 195–197, with an English translation of the relevant passages.
154 Falloppia, Observationes (1561), foll. 27v–28r and fol. 149v; cf. the photographic reproduction and translation in Macchi et alii, Gabriel Falloppius (2014), fig. 1.
155 Falloppia, Observationes (1561), foll. 151v–154v.
156 Ibid., foll. 155r.
157 Achillini, Annotationes (1520), fol. 13v: "Est aliud par nervorum subtile exiens a posteriori cerebri, transiens ad anteriora super loco aurium: de quo nihil inveni a doctoribus puto quod det motum superciliis et statim apparet cum elevatur cerebrum."
158 Falloppia, Observationes (1561), foll. 155r–156r.
159 Ibid., foll. 64r–66r and foll. 68r–71r; cf. O'Malley, Gabriele Falloppia's account of the orbital muscles (1968), with an English translation of the relevant passages.
160 Vesalius, Fabrica (1543), pp. 240–241, illustration on page 239.
161 Colombo, De re anatomica (1559), p. 124.
162 ÖNBW, Cod. 11210, foll. 28r–v, "In brutis sunt 7, in homine tantum 6." Accordingly, Handsch corrected his earlier notes on the public anatomy under Fracanzano in

1550/51, and changed the number of eye muscles, in a different ink, from "7" to "6" (ibid., fol. 25v).

163 Ibid., fol. 148v.
164 SUBG, Ms. Meibom 20, fol. 199r: "Credo ego igitur esse quasdam fibras in ipsis palpebris quae non possunt auelli, quae sunt causa huius motus sursum."
165 Falloppia, Observationes (1561), fol. 65r.
166 Several decades later, Aranzi claimed that he and his uncle Bartolomeo Maggi had seen this muscle in 1548 already, when Aranzi studied anatomy with Maggi in Bologna. He thus suggested that Maggi had discovered this muscle and rejected the claims of those who prided themselves of its discovery after Maggi's death but no independent evidence is known that would support his claim. Undoubtedly, he was referring to Falloppia (Aranzi, Anatomicae observationes (1587), p. 67):

> vere affirmare possum, mihi adhuc adolescenti, primaque anatomes rudimenta ab anno usque 1548 addiscenti, hunc ipsum musculum perspectum fuisse, cum praeclarissimus anatomicus Bartholomaeus Magius avunculus ac praeceptor meus, tunc publice anatomen Bonon[iae] summa cum laude profiteretur; qua propter satis patet, eos, qui aliquot annis postquam ille ex humanis excesserat, floruere, huius musculi auctores minime censendos esse.

167 Falloppia, Observationes (1561), fol. 70r–v.
168 Gitter, Geschichte der Hörforschung (1990). Politzer, Geschichte (1907).
169 Politzer, Geschichte (1907), pp. 89–94.
170 ÖNBW, Cod. 11210, fol. 25v, later addition.
171 SUBG, Ms. Meibom 20, fol.133v: "In osse petroso est cauitas [...]: huic cauitati adiungitur membranula cui superponitur malleolus, et in parte superiori ossis est incus, cum gemino pede, quorum alter fundatur super ossiculum stapes dictum."
172 Falloppia, Observationes (1561), foll. 25r–26r; see also O'Malley/Clarke, Discovery (1961); Gioffré/Di Pietro, Contributo (1963).
173 Falloppia referred to Berengario da Carpi and Mondino de' Luzzi.
174 Colombo, De re anatomica (1559), pp. 26–27.
175 SUBG, Ms. Meibom 220, fol. 133v.
176 Cf. Franklin, Ductus venosus (1941); Zampieri et alii, Three fetal shunts (2021).
177 Falloppia, Observationes (1561), foll. 130r–131r.
178 In *De usu partium*, Galen devoted only one sentence to a tiny vessel he had found which connected the two large vessels (Kühn IV, p. 244).
179 Falloppia, Observationes (1561), foll. 130r–131r.
180 Ricon Ferraz, João Rodrigues de Castelo Branco (2013), p. 493: "Em 1547, Amato Lusitano descobriu as válvulas das veias, na sequência de inúmeras experiências realizadas em animais e em cadáveres humanos, e demonstra-o publicamente."
181 Amatus, Curationum (1551), pp. 258–260.
182 Falloppia, Observationes (1561), 118v–119r.
183 Vesalius, Examen (1564), p. 83, "mihi retulit, se in vena coniuge carentis initio, et item in uenarum renes adeuntium et in sectionum uenae iuxta elatiorem sacri ossis sedem occurentium orificiis membranas eiusmodi obseruare, quales in venae arterialis et magnae arteriae occurrunt principiis, hasque sanguinis fluxus obstare asseruit"; on the dating of Vesalius' encounter with Canani see also Petrucci, Vite (1833), pp. 94–95.
184 Zaffarini, Scoperte (1909); Franklin, Valves (1927), p. 4.
185 Steudel, Entdeckung (1955).
186 Estienne, De dissectione (1545), pp. 182–183.
187 Ibid., p. 357.

188 Dubois, In Hippocratis (1556), fol. 31r: "Membranae quoque epiphysis est in ore uenae azygi, uasorumque aliorum magnorum saepe, ut iugularium, brachialium, cruralium, trunco cauae ex hepate prosilientis, usus eiusdem cum membranis ora vasorum cordis claudentibus."

189 Falloppia, Observationes (1561), esp. foll. 62v–63r and foll. 71r–79r.

190 Ibid., foll. 85r–91r.

191 Eustachio, Tabulae anatomicae (1714), plate XXXIII with Lancisi's explanation on p. 84.

192 Biblioteca comunale, Siena, Misc. XVI, C IX 17, "Bartholomaei Eustachij [...] Anatomica tractatio ex Vexalio [sic!] et alijs anatomicis", fol. 43r: "in fine vero prope ossa pubis bini parui musculi ab omnibus ignorati conspiciuntur." The hand seems to be that of Eustachio himself.

193 ÖNBW, Cod 11210, fol. 145r:

> Est adhuc unum par ultimum, iuxta finem rectorum, in regione vesicae, et sunt parui, alii reputauerunt appendices esse rectorum, et omiserunt hos duos musculos, sed non sunt appendices, immo sunt distincti a substantia et a fibris rectorum musculorum, et non sunt recti, immo in latus quodammodo inclinantur [...]. Praeterea usus admonet, nam sicut intestina debebant comprimi ad excretionem fecum a suis musculis ita et vesica a suis ad excretionem urinae. Exprimunt igitur urinam e vesica <added: dum moventur versus principium suum sc. versus os pubis nam sic contracti deorsum comprimunt vesicam> et ideo urina retenta (nisi sit obstructio) compressis manu his musculis excernitur.

194 Falloppia, Observationes (1561), fol. 25r.

195 E.g. ibid., fol. 147v, "error Vesalii", on the thoracic muscles.

196 ÖNBW, Cod. 11210, fol. 156r; cf. Vesalius, Fabrica (1543), p. 284.

197 SUBG, Ms. Meibom 20, foll. 132v–133r: "Ex eodem ramo, ex quo axillaris etiam humeraria et cephalica oritur: alio modo ut describit Vesalius & Galenus in homine: at in simia vere oritur secundum descriptionem illorum ab externa jugulari."

198 For example, Falloppia, Compendium (1585) foll. 17v, 18v, 23r, 26r, 41v (twice), 43v, 47r ("Vesalii errores"), 52r ("Vesalii incuria"), 52v, 53v, 59r, 64r and 64v; further passages are marked with "contra Vesalium".

199 Falloppia, Compendium (1571), fol. 26v, on the *musculi intercostales*: "circa hos deceptus fuit Vesalius, cum dixit [...]".

200 Ibid., foll. 20v–21r.

201 Ibid., fol. 6v.

202 SUBG, Ms. Meibom 20, fol. 247r.

203 Vesalius, Examen (1564).

204 Printer's letter to the reader, in: Vesalius, Examen (1564).

205 Vesalius, Examen (1564), p. 2.

206 As Castiglioni, Fallopius (1962), p. 189, bluntly put it: "Vesalius' reply to Fallopius is essentially weak."

207 Vesalius, Examen (1564), p. 171.

208 Ibid., pp. 72–73.

209 Ibid., p. 2 "ut diligentius illas discutiant, et locupletiorem artis cognitionem (ne dicam veluti mei de Humani corporis fabrica operis appendicem) sibi comparent."

210 Vesalius, Examen (1609).

211 Cole, History (1949), pp. 58–59.

212 ÖNBW, Cod. 11210, fol. 24r:

> Hoc plexus retiformis non continetur in homine, sed in equis et bobus est exquisitus, nam cum equus exercendo, currendo & bos trahendo incalescunt,

fit quod multus sanguis venosus et arteriosus sursum ascendit, ex quo ex-
udante & aere per nares inspirato fit spiritus animalis. Si autem ille sanguis
arteriosus in tam multa copia a repente irrueret, possit spiritum animalem
suffocare. Ideo natura fecit hos plexus reticulares ut multus sanguis possint
contineri, sed in pluribus & implicatis vasis, ne tam confertim irruat in ven-
triculos cerebrj.

Similarly SUBG, Ms. Meibom 20, fol. 210v, adding that the head was low when
cattle were pulling carts or ploughs making it easy for the blood and the spirit
to reach the head where the plexus prevented suffocation.

213 ÖNBW, Cod. 11210, fol. 28v.
214 Falloppia, Observationes (1561), fol. 176v.
215 ÖNBW, Cod. 11210, fol. 27v:

Cur reliquia animalia non loquuntur, cum habent similem substantiam lin-
guae omnes musculos & fibras. Responsio: Quia lingua pro articulatione de-
bet esse mediocris non longa ut in canibus, non acuta ut in auibus, non crassa
ut bobus, sed mediocris substantiae in fine. Praeterea animalia non habent
intellectum pro imitatione.

216 Ibid., fol. 197v:

ideo maius est hoc caecum in cattis, muribus, etc., quia illi cum per parietes
descendunt feces possint recidere in gracilia intestina. Sic proni incedunt et
feces facile possint recidere in gracilia intestina. Sic eciam maius est in porcis
quia illi, ut pedes habent breves, sic proni incedunt, et feces facile possint
regurgitare.

217 Ibid., fol. 197v, added note in the margin: "In simia caecum certe est maximum
et multo maius quam in porco quoniam simiae plurimum saltant ab arboribus & saepius concitatissime faciunt impetum in pronam partem, ideo ne feces
regurgitarent ad ventriculum habent caecum maximum." Similarly SUBG, Ms.
Meibom 20, fol. 140r: "Et inter animalia quae magis prona inedunt, ea etiam
maius habent hoc intestinum: in quorum numero simiae, quae perpendendent
ex arboribus."

218 ÖNBW, Cod. 11210, fol. 107v:

At in homine est pauci et fere nullius momenti, quia rectus incedit. At dices
tu, Ad quid ergo natura fecit hoc in intestinum in homine cum natura nihil
faciat frustra? Responsio: Sicut vasa umbilicalia in adultis nullius usus sunt,
at in formando foetu fuerunt necessaria, sic hoc coecum in foetu fuit valde
utile, quia ille effertur per caput ex utero matris, et propterea in infantulis
natis hoc intestinum (proportione habita) multo maius est quam in adultis.

Similarly SUBG, Ms. Meibom 20, fol. 139v, "in adultis nullum habet vsum".
Handsch added a little note, with a different quill and ink, in between the lines
that the part also still had its uses in small children, because they crawled on all
four and "I believe that nature gave it to man most of all so he would not be inferior to other animals" ("Et credo quod natura hoc potissimum dedit homini ne
cederet alijs animalibus"). It is not clear whether Handsch is still quoting Falloppia
here – which would imply a certain contradiction – or expressing his own opinion.

3   Surgery

Falloppia is known today primarily as an anatomist. However, like Vesalius before him, as a professor in Padua, and other leading anatomists of the sixteenth century, he also taught surgery. The famous Spanish surgeon Daza Chacón, who worked with him at the Spanish court, later praised Vesalius' anatomical skills but harshly criticized his blunders as a surgeon, describing how he basically killed a patient when he opened an empyema.[1] Falloppia, by contrast, was one of the most famous and most respected practicing surgeons of his time.

In the contemporary understanding, surgery comprised the treatment of wounds, dislocations, and fractures as well as ulcers, superficial tumors, and other skin conditions, including the French disease. Falloppia's lectures on a wide range of surgical topics have been documented in student notes. They occupy more than half the space in the later editions of his *Opera* and they also dominate the handwritten tradition of student notes on Falloppia's lectures.[2] Clearly, Falloppia's students valued his surgical teaching highly. It had a considerable impact also beyond the narrow circle of learned, Latinate medicine. In 1602/03, Giovanni Pietro Maffei, a surgeon from Treviso, published an Italian translation of Falloppia's surgical lectures.[3] According to the subtitle, the edition was aimed especially at barber-surgeons. In his dedicatory epistle to Federico Cornaro, Maffei explained that he wanted to make Falloppia's work accessible to the experienced surgeons of his homeland and to all other *medici volgari* who did not know Latin. Reprints of 1620, 1637, 1647, and – by three different printers in – 1675 underline the popularity of this vernacular edition.[4] Surviving manuscripts with English translations of various surgical lectures by Falloppia also point to the perceived values of his work for nonacademic, non-Latinate surgeons.[5]

In Italy and France, Lanfranc of Milan (c. 1250–1315), Henri de Mondeville (c. 1260–1325), Guy de Chauliac (c. 1300–1368), and other learned surgeons had already brought surgery closer to learned, scholarly medicine in the Middle Ages, insisting on their "rational" method in contrast to the "empirical" practitioners.[6] They combined an extensive knowledge of ancient and Arabic surgical writings with the knowledge and experience they themselves acquired as practicing surgeons. With his *Chirurgia magna* Guy

DOI: 10.4324/9781003242000-4

de Chauliac created a *summa* of surgery that shaped learned surgery for centuries.[7] Even in Italy and France, however, the surgeons succeeded only to a limited extent in gaining the status and recognition – also in terms of salaries – that theoretical and practical medicine and their representatives enjoyed. Separate chairs of surgery with experienced surgeons did not become important until the sixteenth century. The surgical professorship of Berengario da Carpi in Bologna in 1502 – he was himself the son of a surgeon – was an important first step.[8] In the German-speaking territories north of the Alps, the learned doctors in Falloppia's time still left surgery almost entirely to the barber-surgeons, who acquired their knowledge and manual skills as apprentices and journeymen from experienced masters of the trade or in military campaigns.[9] A man like Falloppia, who was not only well versed in the Latin and Greek surgical literature but also treated numerous surgical cases himself – sometimes in the presence of his students – could undoubtedly exert a lasting effect on those numerous budding physicians who flocked from north of the Alps to the upper Italian universities.

The traditional disdain for surgery within scholarly, academic medicine was ultimately based on its perception as a menial, manual activity. The word "surgery" derives from Greek root for "hand", "χεῖρ", and by definition primarily referred to a work of the hand, not of the mind. In contrast to the primarily intellectual activity of the learned physician, who uncovered the mysterious processes inside the body and treated them with medicines to be taken internally, it was considered inferior.

The realm of surgery in the sixteenth century differed considerably from that today. In the practice of the overwhelming majority of surgeons, invasive operations played only a marginal role or indeed none at all. These interventions were essentially limited to the surgical treatment of cataracts, the removal of bladder stones, and the treatment of hernias. The latter two had a high lethality rate and were often performed by surgeons who specialized in this area. As indicated above, the focus of surgery in everyday practice was on two other areas, which we classify as separate disciplines today. One was traumatology, the treatment of wounds, injuries, and fractures, which is still a core area of surgery today. The other one were pathological changes in or underneath the skin and the mucous membranes, which today would usually be considered the domain of dermatology or, depending on the localization in the body, of stomatology, urology, gynecology, venereology, etc.

This division into two major fields is reflected in Falloppia's surgical teaching. On the one hand, he lectured on wounds, fractures, and other injuries. On the other hand, he discussed preternatural "tumors" or swellings and ulcers, and in the latter context also dealt with the French disease, which was often characterized by boils and ulcers.

In hindsight, the historical significance of Falloppia's surgical work does not lie in any revolutionary new concepts or innovative therapeutic procedures. Older work on the history of surgery, which was primarily interested in the contribution of individual authors to the "progress" of medicine, understandably did not see Falloppia as one of its great icons. Drawing on his extensive knowledge of the older surgical literature, Ernst Gurlt found Falloppia's surgery to be epigonal.[10] Today, Falloppia's surgical work, his advice on how to diagnose and treat a wide range of surgical (and dermatological) conditions may appear outright worthless and except for his various recipes for ointments and other medicines, much of what he had to say was not even new in his time.

Falloppia's importance for the history of surgery lies above all in his role as a trailblazer: Falloppia was a leading exponent of the new humanistic surgery and played a significant role in elevating surgery within learned academic medicine. Surgery had been one of the three pillars of medicine since antiquity, he proclaimed to his students, along with the treatment of diseases with medicines and dietetics. A good physician also had to master surgery and the true surgeon, in turn, could not want to be just a surgeon: he had to master medicine in its entire scope.[11] This was the lesson which his numerous students from north of the Alps also took back home.

Falloppia – and this explains Gurlt's negative judgment – relied largely on the extant surgical literature. He quoted widely from older and more recent surgical writings and strongly recommended that his students read the relevant passages before his lecture on a certain topic, which would also allow them to see where he deviated from the teachings of the ancients. He referred his students in particular to the Hippocratic writings on head wounds and fractures, the surgical parts of Galen's *De methodo medendi*, the seventh and eighth books of Celsus' *De medicina*, the sixth book of Paul of Aegina's compendium of medicine, and the works of the Arabic writers. He also quoted contemporary writers like Guido Guidi, who taught in Pisa.[12] His principal authority, however, was Guy de Chauliac, whose work he recommended reading in the Latin translation by Jean Tagault.[13]

Falloppia was not only one of the most famous representatives of humanist surgery.[14] In his teaching, he also brought a major new trend in medical education to bear on the teaching of surgery, namely the rise of bedside teaching.[15] Students at Padua, as at other leading universities of the time, sought, demanded, and found an education that prepared them comprehensively for the treatment of patients in everyday practice, from which the vast majority of them would later have to earn their living. Falloppia's lectures were decidedly oriented toward surgical practice. He repeatedly described to his students, down to the smallest detail, how they had to proceed in the treatment of various surgical ailments. He presented the different instruments and their use, for example, whether it was better to use a straight or a curved knife for certain surgical activities.[16] He gave his students countless recipes for external and sometimes also internal remedies, frequently adding which of these he preferred ("ego autem soleo") and which ones he used sometimes only ("aliquando utor"). And he not only referred to his good personal experiences ("cum magno successu") in general but also cited concrete cases in which he had used them successfully on a sick or injured patient.

Falloppia's surgical teaching, moreover, was not limited to his lectures. Falloppia repeatedly used anatomical demonstrations to instruct the students about surgical matters. For example, he showed different techniques that could be used to suture abdominal wounds, from the rather coarse suture of the kind that fur makers used to the more refined technique of joining the different layers of the abdominal wall separately. The Helmstedt Anonymus even added little drawings that illustrated the various types of sutures. Apparently, Falloppia demonstrated them on a corpse.[17] Falloppia also taught his students how to perform a paracentesis to relieve dropsical patients of some of the massive amounts of water that collected in their abdomen. He showed them where they had to cut through the abdominal wall, namely on the side and below the navel about three fingers from the hipbone. Then they had to insert a tube through which the fluid could drain off. Drawing on his anatomical knowledge, he also warned them of the dangers, however: they would have to cut through the peritoneum, which clothed the walls of the abdominal cavity and this carried great risks. He had performed the operation himself on three patients and they all had sooner or later died. One should therefore only perform a paracentesis at the patient's request.[18] Falloppia moreover showed his students, on the corpse, the passage of different vessels and the corresponding openings in the groins, which allowed the intestines or the omentum to pass through in cases of inguinal or scrotal hernia.[19] The student on whose notes the *Compendium* was based likewise documented Falloppia's references to surgical interventions, such as his warning that cutting through the sphincter of the bladder during the surgical removal of a bladder stone could easily cause urinary incontinence after the operation.[20] Accordingly, Falloppia also referred to his anatomical demonstrations in his surgical lectures. They should recall what they had seen in the public anatomy, he admonished his students at the beginning of a series of lectures on wounds.

Last but not least, and this was a central element of the practical surgical training that Falloppia offered them; the students were allowed to accompany him on his visits to the sick. At the beginning of his lectures, he repeatedly promised that he would take them with him to see patients during their free time. This would be of the greatest benefit to them. For some things were so difficult and complicated that they could only be learned at the bedside.[21] This was not just an empty promise. In his lectures, he sometimes explicitly referred to concrete, individual patients, sometimes even giving their names, whom his listeners – or at least some of them – had seen together with him.

In the following, I will give a brief overview of Falloppia's teaching on the various fields of surgery, focusing on the major topics and highlighting those aspects, in particular, that were of interest to the practical training of students or where Falloppia contributed new or differing opinions, personal assessments, and practical experiences.

Dermatology: ulcers and other skin conditions

The very first among the various works by Falloppia that came out after his death was devoted to ulcers and superficial tumors or swellings. The Venetian printer Donato Bertelli published it in 1563 under the title *Libelli duo alter de ulceribus, alter de tumoribus praeter naturam*.[22] It was based on a set of student notes. Bertelli claimed that he had paid a considerable amount of money for these notes but did not reveal his source. Maybe he owed them to Petrus Angelus Agathus who, a few months later, published his notes on Falloppia's lectures *De morbo gallico* with Luca Bertelli.[23]

As the title indicates, the publication consisted of two parts, dealing with ulcers and tumors, respectively. In 1577, Bruno Seidel published his own, much more extensive notes on Falloppia's lectures on ulcers he had attended in 1557/58 under the title *De ulceribus liber*.[24] It included a number of chapters on types of ulcers that were missing in Bertelli's edition. An unidentified student also took notes on Falloppia's lectures on ulcers in 1560/61.[25] In the following account of Falloppia's lecture *De ulceribus*, I will take the relevant section of the original publication of 1563[26] as my starting point and draw on Seidel's 1577 edition for the topics it does not cover.[27]

The lecture focused on conditions which would mostly be assigned to the domain of dermatology today. Falloppia even discussed skin changes that did not result in ulcers at all. He began with a general discussion of the art of surgery of the "ethimologia" of the term "chirurgia", which – as his students thanks to their training in Latin and Greek most likely already new – quite simply meant "manual work" ("manuaria opera"). The practice of surgery, as Hippocrates and Celsus had already underlined, demanded a firm hand, a sharp eye, and a mind that was not easily shaken and calm compassion, avoiding to inflict unnecessary pain on his patients. Falloppia also described the suitable conditions for surgical practice: good light, a comfortable position for the surgeon (when the surgeon was sitting, he should sit straight, with his knees about the width of an arm apart and his hands should not need to reach higher than his breast). He had to wear simple, suitable clothes and his fingers should not be full of rings. His fingernails should not reach beyond the fingertips.[28] He also described – and presumably showed to his students – a range of instruments which the surgeon needed for his work: scalpels and knives of different shapes and sizes, a saw, different types of forceps, needles, special tools for pulling teeth, extracting arrows and bullets or removing polyps, cautering irons, and the like. Interestingly, in the light of widespread complaints about female shame that made diagnosis difficult, he also listed a small clyster for vaginal application and a speculum "pro uteri fundo inspiciendo", suggesting that some female patients at least did permit male surgeons to see and examine their private parts – and that his students would need these tools.[29]

Turning to the more specific topic of his lecture, Falloppia discussed the causes of ulcers, in general. Ulcers developed when evil, corrosive humors or fluids ("juices") flowed toward the skin. These "juices" resulted from a *cacochymia* of the body or one of the major organs. There also were excrements from individual parts of the body, however, he added, and fluids that originated in the affected part itself, as in the case of frostbites. Certain ulcers were ultimately due to contagion, which was transmitted by vapors or direct contact. Thus, lung ulcers resulted from foul vapors, which were "infected" with putrefaction. Just sharing the bed with someone who suffered from scabies could lead to scabies. Sexual intercourse with a woman who had a genital ulcer would cause an ulcer also in the male member.[30]

As Falloppia explained, some ulcers (and other skin conditions) only affected the "decoratio". They were a cosmetic, aesthetic problem but did not impair function and thus did not fulfill the accepted Galenic criterion for a disease. In other cases, pain or itching (which according to Falloppia was also a kind of pain) affected the action of the sensory faculties, however, and therefore constituted a disease.

The treatment of ulcers followed three indications: (1) drying and/or evacuating the corrosive humor, (2) tightening and strengthening the affected part, and (3) fighting the cause of the humoral flux into it. Among the remedies that could be used, Falloppia recommended in particular, waters that contained alum, like those in Abano. They had a drying, cleansing, and repelling effect when the ulcers were washed with them and could also be prepared artificially. In some cases, it was moreover necessary to promote the regeneration of flesh and skin.[31]

After this general introduction and overview, the remainder of the lecture dealt with the different types and localizations of ulcers in the body and their treatment as well as with other changes in the skin, such as scabies, *psora*, impetigo, and leprosy. Seidel's 1577 edition expanded that range considerably, with sections on special subtypes such as *ulcus phagedaenicum* and *ulcus nomosum*. Seidel's notes included chapters on ulcers according to their anatomical location, such as ulcers of the eyes, the nose, and the gums. Falloppia also dealt in detail with cancer, which according to the contemporary perception[32] manifested itself primarily as a cancerous ulcer.[33] We will return to this topic in the context of Falloppia's approach to a "palliative cure".

The lecture, as documented by Seidel, concluded with a detailed discussion of the diagnosis and treatment of anal fistulae, which offers another example of the very much practice-centered approach Falloppia took in these lectures.[34] Falloppia explained how the surgeon could diagnose and assess the precise location and extent of a patient's anal fistula. When the fistula had an opening to the outside, urine, feces, or pus might exit visibly through the opening. When some fluid was injected into the fistula with a clyster some of it would find its way into the anus, and when the surgeon put some fat or oil on his finger, inserted it into the anus and pressed in the direction of the fistula, some pus might come out. Fistulae that did not open toward the outside could be sounded with a probe ("specillum"). Many fistulae were not straight, however. They followed a curved, tortuous trajectory. A lead or tin probe, as some surgeons used it, was not suitable in such cases. Falloppia recommended the use of a little stick made of white wax instead, which could be pushed forward more smoothly.

The cure of anal fistulae was often difficult and sometimes impossible, especially when the fistula excreted urine or extended into the rectum and thus was in constant contact with the excrements, which did not allow the opening to heal. Even when the fistula was curable, Falloppia recommended a palliative treatment, a "curatio paleans": except for very mild cases, it was preferable to leave the fistula open and just make sure it did not become larger and tell the patients that it was better that they live with if they could. For Nature used the fistulae as a means to evacuate excrements and he had seen serious harm that resulted from making them close.

If the patients asked for a definitive cure, the body first had be cleansed. The fistula itself was widened, the hardened parts ("callus") was removed by injecting *radix gentiana* or similar medicines, and then closing the fistula could be tried. If this treatment was not successful, a cure could be achieved by cutting. A special surgical instrument which was equipped with a blade was to be used for this purpose. The way it was used could only be taught on the patient. It was introduced into the fistula and pulled out through the anus. With the alternative method – many patients did not tolerate the use of the instrument – which was also already described by Hippocrates, a wire was pulled through the fistula with the help of a probe. Over the course of several days, the wire was gradually pulled toward the outside. This could only safely be done, however, when the opening of the fistula was not too far away from the anal opening. The sphincter muscle was four-finger-widths long. If the instrument or the wire cut through no more than two- or three-finger-widths, it would still preserve its function and the patient could control his stools. But if more of the sphincter muscle was cut through, the patient would become incontinent and the surgeon must never cut so deep.[35]

Caustics and cauterization

Cauterization, the deliberate creation of an ulcer by means of heat or caustic substances, was a commonly applied method in the borderland between surgery and internal medicine. Student notes on a lecture by Falloppia on this subject were first published under the title *De cauteriis* in 1570, as an appendix to his lectures *De compositione medicamentorum*.[36] Since his remarks on the topic were quite brief, Falloppia probably dealt with the topic in the context of another lecture, and more precisely in the context of ulcers: Seidel published Falloppia's detailed discussion of cauterization as a part of his 1577 edition of his notes on Falloppia's lecture on *De ulceribus*.[37] He also promised his students that he would teach them the application of the cauter on actual cases.[38]

The term "cauter" was often used at the time to refer to an ulcer that was created in the arm, thigh, back, or neck in order to divert the flux of morbid matter, which otherwise threatened to accumulate in the diseased part of the body. This was commonly also called a "fontanel", Falloppia explained, but it was not what "cauter" originally meant to the ancients. The techniques used in cauterization and in creating a fontanel were largely the same but cauterization in the strict sense of the word was used to treat directly the sick part of the body.[39]

Two basic forms of cauterization had to be distinguished, depending on whether the heat applied to the skin acted "potentialiter" or "in actu", the latter referring to heat that could be directly perceived with the senses. In his description of cauterization by heat "in actu", with a hot iron, Falloppia could limit himself to a few words. It was a commonly known procedure and widely practiced despite the pain. He only briefly discussed the suitability of other metals than iron for this purpose and described the (rarely practiced) cauterization with boiling water or oil or with liquid sulfur or lead. Here one had to ensure that the effect remained strictly confined by a wall of clay or a similar matter around the site of application.

The bulk of the tract was devoted to a series of corrosive agents of varying strength that acted "potentialiter". Their effect resulted from a "potential", hidden heat that was not immediately perceptible to the senses. The spectrum of suitable substances ranged from cantharides and alum to the powerful *aqua sulphuris*, also called *oleum vitrioli*, essentially what we would call sulfuric acid today. In this context, Falloppia also described in detail how the "sulfur water" could be made in a kind of distilling furnace as it was used by the "chymistae". The handwritten notes of a student on Falloppia's lecture, probably by Theodor Zwinger (1533–1588), who was studying in Padua at the time, suggest with various sketches that Falloppia even concretely demonstrated to his audience the chymical apparatus that was used for this purpose or at least illustrated the construction with images the students could copy.[40]

Falloppia summarized his lecture in a dichotomous, branching table, of the kind which enjoyed considerable popularity at the time. It extended over several pages and allowed his students to grasp at a glance the various animal, vegetable and mineral substances, simple and composite, which could be used for cauterization, with the relative intensity of their heat and their other qualities.[41]

The French disease

The French disease or *morbus gallicus* caused great concern among physicians and laypersons at the time.[42] It was widely perceived as a new disease and it raged with devastating consequences. Whether the disease was caused by the same pathogen as syphilis today is still open to debate but contemporary descriptions certainly suggest that the symptoms and the long-term sequels of the disease were often more serious and dramatic than those associated with syphilis today. Many patients suffered from excruciating pain, especially in their bones. Their bodies were covered with boils and festering sores. The stench of putrefaction could become unbearable. With time, many also lost their hair and their nose or other parts of their face might be literally eaten away as well.

From the late fifteenth century, when the first reports circulated, numerous authors dealt with the diagnosis and treatment of the new disease. By Falloppia's time, the disease had long become a topic of academic lectures. Against this background, the lectures of a famous and experienced learned surgeon like Falloppia were bound to meet with great interest. Falloppia's lectures on the French disease have survived in printed and handwritten student notes, in Latin and in vernacular translations.[43] Already in 1563, a year after Falloppia's death, Luca Bertelli in Venice published the notes of Petrus Angelus Agathus on Falloppia's lectures under the title *De morbo gallico liber absolutissimus*, together with the lectures of Falloppia's former colleague Antonio Fracanzano, who taught in Bologna for some time.[44] In 1565, a second edition was published by F. L. de Turino in Venice.[45] In 1574, Agathus' notes came out in two other editions, with the heirs of Melchior Sessa and with Aegidius Regazola.[46] The Frankfurt editions of 1585 and 1600 reprinted the text, with only minor editorial changes.[47] The Venetian edition of the collected works of 1606, on the other hand, published the notes of another student – according to the publishers they had acquired the notes of Andrea Marcolino – on Falloppia's lectures on the ulcers, which included a detailed discussion of the French the disease.[48] The two versions, by Agathus and by Marcolino, follow exactly the same structure and a rough comparison reveals no fundamental differences in content. Those of Marcolino are more detailed overall than those of Agathus but some passages are also more succinct. Falloppia read repeatedly on ulcers and the French disease – as far as can be discerned every three years, in 1554/55,[49] 1557/58,[50] and 1560/61[51] – and probably Agathus and Marcolino heard the lectures at different times.

Although he mentioned the name "syphilis" coined by Girolamo Fracastoro,[52] Falloppia like most contemporary authors outside of France called the disease "morbus gallicus", the "French disease". Falloppia discussed the disease in a great detail and took a position on the various questions discussed in the medical literature of the time.[53] He had no doubt that it was a new disease.[54] It had been sent as a divine punishment for the moral decay of mankind but it was undoubtedly transmitted by contagion. The Spaniards had brought it to Europe, in 1494, when they returned to Europe from the West Indies with Christopher Columbus. While the disease was mild in the West Indies, similar to scabies, it took on a savage, merciless character when it reached Europe.[55] The first ones to be infected were the French troops who besieged Naples in the spring of 1495. Falloppia – who repeatedly showed a certain dislike for the Spaniards – thought that the Spaniards in the besieged city had spread the disease among the French on purpose. They had not only poisoned the fountains and bribed the Italians bakers so they would add plaster to the bread they delivered to the French. They also banished the most beautiful prostitutes from the city.

Not surprisingly, the French, who liked women, took them in and took pleasure in them without restraint and so the whole French army and finally the whole of Europe was infected.[56]

Falloppia rejected the claims of some authors that the disease had already been described by the ancients or by the Arabs. Those who attributed the disease to an unfavorable planetary constellation, in October 1483, in the eighth house (of the twelve houses of nativity) which, according to the astrologers signified health, went even further astray.[57] In 1550s' Padua, astrology had its official place in the arts faculty. Pietro Cadena (or Catena) lectured on the topic.[58] Falloppia, however, had little sympathy for astrology. The celestial bodies, he explained, had an effect on the earthly events by their movement and their light only. The aspects of the planets had no influence and the physician must not attach any importance to them. Even if one were not to reject astrology per se, he added, the question remained why this allegedly "unfavorable" planetary constellation would unfold its effect only eleven years later.

Falloppia concurred with the opinion of other writers that the disease was mostly but not always transmitted by sexual intercourse. He did not believe that it could be transmitted by drinking from the same glass or cup[59] but in some cases just kissing, spending a night in the same bed, or wearing the clothes of someone with the disease were enough.[60] Moreover, he had seen infants, who were born with French disease, and infants could become infected when they drank from the breast of woman who had the disease. From all this, Falloppia concluded that the principal location of the disease in the body were not the genitals. It often first manifested itself in the genitals but neither the genitals nor the skin were the true subject of the disease. The *contagium* affected primarily the liver.

As with various poisons and pestilential fevers, the devastating effects of the tiny, invisible *contagium* did not come from a peculiar mixture of humors or quality, in Falloppia's opinion. It sprang from its special nature, its total substance.[61] It had a special affinity to the liver as the seat of the natural faculty and to its principal tool, the natural spirits. When the *contagium* reached someone else's body through the secretions from an ulcer or through the vapors affected patients released through the pores of the skin, it reached the liver through the vessels. This happened all the more easily when the vessels in the "recipient" body were wide open – for example, due to the heat of sexual intercourse. From the liver, the *contagium* reached the genitals, the skin, and all the other parts of the body with the blood, and the harmful vapors the sick body released were, in turn, passed on to others.

The central role of the liver and the natural spirits in the disease process helped explain the multitude of possible symptoms and sequelae, which by no means always appeared together in the same patient. Moreover, the disease had different phases ("tempora"). Immediately after the infection, the patients suffered from a certain heaviness and fatigue, sometimes also from pains with a changing localization; this was due to the affection of the natural spirits. The color of the face changed as well. Sometimes livid circles appeared under the eyes, as they could be seen in women during their menstruation.[62] As a "signum patocnomicum [sic!]",[63] a genital ulcer usually formed soon after. It permitted the diagnosis of *morbus gallicus*, if its appearance was preceded by a suitable opportunity for the transmission of the *contagium*. Soon a swelling, a bubo, developed in the groin, and "gonorrhoea" set in, that is, literally a flow of semen: due to the weakening of the natural spirits, according to Falloppia, the testicles could no longer retain the semen and it constantly dripped.

At a later stage, some four to six months after the infection, hard pustules appeared all over the body, some with crusts, others with oozing secretions, and malignant ulcers developed. The ulcers on the genitals hardened. The throat, palate, and uvula were affected. The voice became rough. Tumors formed the so-called gummata (as they are still called today). The bones were attacked. Painful fissures and crusts appeared on the palms of the hands and feet. Falloppia thought – and this is quite possible – that he had been the first to discover another characteristic symptom, tinnitus, which had not been described by anyone before (and which is known today as a typical symptom of advanced syphilis). It was rarely absent in a full-blown *morbus gallicus*, he found.[64]

According to Falloppia, the predominant clinical picture had changed, by his time, since the first appearance in 1494. The typical hair loss, he explained to his students, had only been observed for the last thirty years approximately, and the *gonorrhoea gallica* only for fifteen. Therefore, further changes and the appearance of new symptoms could be expected in the future. At the same time, the power of the *contagium* had diminished.[65]

For students, and thus for Falloppia's mostly young and unmarried male audience, the disease was also of considerable personal relevance. Students were regarded as particularly vulnerable because of their loose sexual mores and their commerce with prostitutes. Falloppia repeatedly wove into his lectures stories of students who used the services of a prostitute and became infected. Thus, he told of seven students who together "appointed" ("conducta est") a prostitute, as that was commonly called according to Falloppia, and they all had intercourse with her. Before doing so, they asked her if she had skin lesions or an ulcer, such as it occurred in the French disease. She denied but subsequently they all developed a typical ulcer. Outraged, they went back to the woman, inspected her in bright light, and saw that her genitals were massively afflicted in fact. In revenge, they applied gunpowder to the open areas and set them on fire, causing the "poor prostitute" to suffer burns.[66] On another occasion, according to Falloppia's account, as many as twelve students got together and took a prostitute into a house. Although some of them apparently had multiple intercourses with her, not all were infected but some got the disease, including one who had intercourse with her only once.[67]

The latter example also served Falloppia as evidence that not every sexual intercourse with an infected man or woman led to disease. Different factors promoted, impeded, or prevented infection. Firm genitals and a *glans penis* that was devoid of a foreskin made it more difficult for the contagium to penetrate through the skin. For this reason, in Falloppia's experience, circumcised men contracted the disease in rare exceptions only. Those who ejaculated quickly were less at risk than those whose intercourse lasted longer. Weakened natural spirits, wide vessels, and a hotter habitus made it easier for the contagion to damage the natural spirits and to penetrate the liver. Old people were colder by nature and therefore less likely to get infected. Some old men had intercourse with prostitutes who had the disease but did not catch it themselves.[68]

Falloppia also attributed great importance to the emotions, the "affectus animi" as they were often called at the time. Those who were inflamed with love for a woman were at a particular risk. Married men and women, on the other hand, often did not infect their spouses. During intercourse with her lover, the adulteress became inflamed with passion and was thus easily infected. When she slept with her unloved husband, on the other hand, she lay coldly and demurely, and was possibly afflicted with a bad conscience because of her adultery – and did not become sick. Therefore, many thought that their wives were respectable and chaste, although, in reality, they suffered from the French disease. In addition, habit had a weakening effect on the passion in conjugal intercourse, so that in those who had been married for a long time and had children, infection was rare. A newly married, loving husband infected with the disease, on the other hand, almost inevitably transmitted the disease to his wife.[69]

Falloppia's discussion of the treatment of the French disease and the various symptoms and complaints it produced was even more detailed. As a general rule, the treatment had to be preceded and supported by measures and means that emptied the excrements that accumulated in the body under the effect of the contagion and the weakened natural faculty via sweat, stool, and other pathways. In addition, it was necessary to act against the heat, to correct the hot and dry *intemperies* that remained in the body even after the cure of the disease itself. However, decisive for a successful cure was a treatment that acted against the nature of the disease as such, that is, against its total substance. Accordingly, a cure could only be achieved with a remedy that, in turn, acted by force of its own total substance.[70] Falloppia discussed among others the use of salsaparilla and cinchona bark but he praised above all the use of guaiac as a true antidote that restored the liver. It could be given as a decoction or in other forms. The treatment had to be supported by dietary measures. In particular, the patient had to eat sparingly, be sexually abstinent, and avoid strong negative emotions. Playing cards, in particular, was dangerous and had to be forbidden.[71]

Falloppia also discussed the treatment with mercury preparations, as they were widely used against French disease at that time, especially by barber-surgeons and non-academic "empirics". To fumigate patients with mercury vapors, he explained, a vessel with glowing coals was placed in a tub, and a kind of pavilion or tent of dense fabric was stretched over the tub to retain the smoke. Then the patient was placed naked on a small stool in the tub and the mercury preparation was sprinkled on coals. Depending on how long the patient could tolerate them, he or she was exposed to the mercury vapors for a quarter of an hour and up to an hour. Afterward, the patients had to lie in bed wrapped tightly and warmly for two hours to promote sweating. The procedure was repeated on the following two days, in the morning, for three cycles of three days each, so that the treatment included a total of nine sessions, sometimes more. If ulcers formed on the palate or diarrhea set in, the fumigation had to be stopped immediately. During the sessions, the doctor had to talk to the patient to make sure that the patient did not lose consciousness. There was a particular danger if the patient also inhaled the vapors over an extended period of time. Falloppia therefore had his patients inhale the vapors only intermittently, just long enough to allow the vapors to enter the body also through the nose and mouth. He gave the patients a long, hollow tube that extended out of the pavilion or tent, which they could use to inhale air from outside. Alternatively, an opening could be made in the tent-like cover so they could put nose through it.[72]

In general, Falloppia advised great caution with mercury. Mercury fumigations were to be used only when at least one of two conditions was met: when all other treatment attempts failed to achieve the desired success or with people in power who had very important tasks and business and feared the shame and infamy of losing their hair due to the disease. Otherwise, mercury fumigations were to be avoided like the devil shied away from the crucifix.[73]

In further chapters, Falloppia dealt in detail with the often very difficult treatment of the various symptoms and described some of these symptoms more exactly: the loss of hair from the head, the beard, and the eyebrows, the characteristic *gummata*, which formed on the head and in the extremities, the pustules on the whole body and particularly in the face, on the hands and feet, and on the buttocks.

The "condom"

One passage in Falloppia's *De morbo gallico* has attracted particular attention in historical writing. At some point, Falloppia described a device to protect oneself from infection with the disease, which he had devised.[74] Based on this passage, 1563 (or 1564, due to an erroneous dating of the first edition) is widely considered to be the year of the first description of a condom, in history, and Falloppia has been praised as its inventor.[75] As manuscript student notes on Falloppia's (repeated) lectures on the French disease show, Falloppia made his invention public earlier already, at the latest in 1555. He therefore must have developed this device already in the late 1540s or early 1550s.[76] And he clearly believed in its value, continuing to teach it until his death.[77]

According to his students, Falloppia addressed them here directly as the potential users and beneficiaries. He wanted to teach them how someone who saw a beautiful "siren" and had intercourse with her could protect himself from *morbus gallicus*, even if she was infected ("infecta"). Since the ulcer was caused by a *contagium*, by impure corpuscles ("corpuscula saniosa"), it was important to clean the *glans* immediately after the intercourse. If the matter had already penetrated into the pores, however, as often happened with a soft, uncircumcised ("tectis") glans, the *sanies* could not be completely removed with wine, water, or any other liquid. Falloppia had therefore searched for a remedy that could render the remaining contagious matter harmless by drawing it out, drying it, or neutralizing its harmful effects. Since this would offend the women – according to some of the notes Falloppia spoke openly of "whores" ("meretrices") – his male listeners could hardly bring vessels with ointments with them; presumably, Falloppia meant that this would have insinuated that the students believed the women were infected. Therefore, Falloppia had, in his own words "invented" ("inueni") a little piece of cloth ("linteolum"), previously soaked with suitable substances and then dried, which they could comfortably carry with them. The liquid with which to soak the cloth was made from various vegetable and animal substances, including guaiacum, which were left to stand for a while and then boiled. A clean piece of fabric was then soaked in this decoction overnight, taken out again, and left to dry in the shade. This process had to be repeated twice. Finally, the piece of cloth was cut to size so it fit over one's *glans* and could easily carried around, in a pocket or pouch.

After intercourse, one had to wash the penis, if possible, or wipe it with a cloth. Then one had to put this *linteolum* over the glans and pull the foreskin over it.[78] It was possible to moisten the piece of cloth beforehand with spit or urine but this was not necessary.[79] If one was worried that an ulcer would develop in the urethra, one could also stuff some of the fabric into its orifice. One then had to leave the piece of cloth in place, on the glans, for four or five hours. If his listeners were looking for a stronger remedy or felt an itch on the genitals, they could moreover resort to fumigations – presumably he meant mercury fumigations – when they returned home.

A closer reading of the passage, which has come down to us, with minor variations, in the notes of different students, leaves no doubt. One can speak of a precursor of the modern condom only insofar as Falloppia's *linteolum*, like the later condom, was adapted to the size of the glans and was put over it. However, it was designed for use after coitus, not during it. It was a preventive, post-coital treatment against syphilis not a contraceptive.[80]

A major reason for the common misinterpretation seems to be the form "coiverit", which Falloppia used twice: "Quoties ergo quis coiverit…" and "cum coiverit ponat supra glandem". The form "coiverit", which we also find in manuscript student notes,[81] is a future perfect.[82] The passages just quoted are thus to be translated literally as "hence as often as someone will have had intercourse" and "when [someone] has had intercourse, he should put it over the glans". In other words, the application of the *linteolum* had to be done every time after one had had a risky intercourse. The translation, "as often as a man has intercourse",[83] is imprecise and misleading.

This is also clear from the way in which Falloppia described the goal of his invention. His explicit aim was to devise a means to pull out or render innocuous the dangerous, impure corpuscula that might already have penetrated through the skin of the glans during intercourse. The requirement, to apply the linteolum over four to five hours, might still be taken to mean only that it should remain on the glans also after coitus. However, the advice that one could also insert a piece of cloth into the urethra is definitely not compatible with an application during intercourse. And in practical terms, it seems impossible that a little *linteolum*, which only covered the glans, would remain in place during intercourse, when it was not held in place – as afterward – by the foreskin; there is no mention anywhere of a ribbon or some other means to tie it to the penis. Moreover, if the women were not allowed to see the ointments men brought with them, it would be hard to understand why a *linteolum*, which was applied before intercourse, would cause less offence, all the more so if it was soaked, before their eyes, with spittle or even urine.

Falloppia claimed that he had tried ("feci experimentum") his *linteolum* on "a hundred and a thousand men" and none of them got infected.[84] This claim has sometimes been taken seriously and Falloppia has been praised for having performed a very early large-scale clinical trial. Some authors have even added up the numbers to arrive at altogether 1,100 men. As with many ancient and Renaissance physicians in this pre-quantitative medical world, such figures must by no means be taken literally, however. They simply stand for a "large" number. Moreover, in this specific case, Falloppia would seem extremely unlikely to have any concrete clinical evidence at all. Maybe some students reported that they had used his *linteolum* and did not develop an ulcer. Since the students surely would not have sexual intercourse with a woman they knew to be infected, they had no means to assess, however, whether they were exposed to the contagion and saved from it by the *linteolum*.

In sum, Falloppia did "invent" a device, a little piece of fabric, which was adapted to the size of the male glans and had to be "pulled" over it. In this sense, his *linteolum* was a precursor of the "condom", although Falloppia did not use that term, which is of uncertain later origin. However, it was to be used after and not during intercourse and would indeed have been unsuitable for that purpose. Certainly, the step toward a device that covered the whole penis and could be used during intercourse would not seem a big one. There is no evidence, however, that the condom, as we know it today, originated from Falloppia's invention. The idea may well have different, independent origins. The crucial issue was probably not the shape but finding a suitable material – such as thoroughly cleansed animal intestines – that would not interfere as much with sexual pleasure or even outright hurt as a sheath made of fabric.

Bumps, swellings, and tumors

Bumps, swellings, and other "tumores praeter naturam" were the third major field of surgery to which Falloppia repeatedly devoted a series of lectures. Our current understanding of "tumors" does justice only to a very limited extent to the broad spectrum of clinical pictures and changes that Falloppia – like other authors at the time – dealt with under the rubric "tumors". The term encompassed all kinds of pathological swellings or elevations on the surface of the skin in which the skin itself remained intact. As in the case of ulcers, many of these pathological changes would today be considered as dermatological disorders. The dividing line between ulcers and tumors was moreover blurred to some degree because swellings or bumps that were initially below the skin surface sometimes broke through the skin and ulcerated.

Falloppia's lectures on preternatural tumors were first published by Bertelli in Venice in 1563. The edition was by all appearances based on an incomplete set of student notes. According to the notes, Falloppia offered an extensive discussion of the doctrine of tumors and their treatment in general. He then, however, only went through some of the most important more specific types of tumors, such as gangrene, carbuncle, bubo, erysipelas, edema, and scirrhi or hardened tumors.[85] In comparison, the notes on Falloppia's lecture on this subject, which were later published in his collected works – presumably based on Marcolino's manuscripts – are more comprehensive.[86] In addition to the topics already mentioned, Falloppia also dealt with atheromata, strumae, leprosy, cancer, tumors of the nose, excrescences on the gingivae, panaritia, frostbites, corns, warts, hemorrhoids, "carunculae" in the urethrae, and swellings of the genital "glands". Even hydrocephalus, inguinal hernia, and a protrusion of the umbilicus in newborns and pregnant women ranked among the "tumors" Falloppia discussed.

The focus of Falloppia's lectures on the tumors was very much on treatment, mostly with a wide range of medicines. He equipped his students with numerous recipes and instructed them on the use of surgical interventions, when necessary. He explained, for example, how polyps could be removed from the nasopharyngeal space by means of a snare. A piece of brass or steel wire was "folded" in the middle to form a loop and both ends were put through a silver tube. The surgeon then had to sling the loop around the base of the polyp, push the tube as far as possible toward the polyp and pull the loop back into the tube to cut through the soft flesh.[87]

Experience and manual skills were very important in this field, Falloppia admonished his students. Even if he just opened an abscess, the surgeon had to know the anatomy, the location of arteries, veins, muscles, and nerves in the area in question. He needed to be familiar with the orientation of the fibers in the muscles because, in most cases, the incision was best made in the longitudinal direction, parallel to the fibers. He had to be prepared for bleeding that might need to be stopped. He had to have the appropriate instruments at hand, differently shaped knives, which Falloppia described and presumably showed. Last but not least, it was important that the surgeon had a sure hand and showed a cheerful disposition, yet, at the same time, acted quickly. It was precisely in this, Falloppia underlined, that the excellent, outstanding surgeon differed from the crude and inexperienced one.[88]

Falloppia described some surgical procedures in considerable detail. In cases of gangrene, for example, an amputation might be called for. They should only perform the operation, however, if the patient and his relatives agreed, or indeed repeatedly asked for it, he warned his students. They had to point out the dangers. A fatal outcome was not uncommon, due to the putrefaction that was communicated to the heart. Before the operation, the patient had to drink two eggs and take some Malvasia wine to strengthen the spirits, because many patients fainted. The sick person's head was best covered so that the patient could not see the instruments. Few were as brave as Giovanni de' Medici, who in Mantua was hit in the leg by a musket bullet and developed gangrene ("sphacela"). Between his screams, he even instructed the surgeon on where to cut and afterward he had the severed limb shown to him. With a red-hot knife – which at the same time had a cauterizing effect – the flesh of the ligated limb had to be cut through to the bone; against Celsus, who had recommended cutting in the healthy part, Falloppia, quoting Galen, advocated making the cut in the gangrenous part. Then the bone had to be sawed through and the bleeding from the large vessels stopped with the cautery iron. After that, the ligature could be released. The bone was cauterized and the wound dressed.[89]

Turning to the surgical treatment of hernias – next to the surgical removal of bladder stones the most important invasive surgical procedure at that time – Falloppia again warned of the considerable risks. In debilitated and elderly patients as in those who were coughing, it should be avoided from the outset. If surgery remained the only option, different approaches could be taken. The groin could be opened with a knife and the vessels leading from there to the testicle could be removed along with the testicle itself. Then the opening in the groin through which the vessels had passed could be sutured. The disadvantage was that removing the testicle could affect the ability to reproduce. Some men were known to have fathered children, however, although they only had one testicle. Other surgeons used a hot iron to cauterize the inguinal region after the intestines had been pushed back into the abdominal cavity, so that they could no longer prolapse. In France, Falloppia related, caustic substances were used for the same purpose.[90]

Palliative care

Falloppia offered his students detailed and remarkably precise instructions on how to treat ulcers and certain tumors with the knife or the cauter when necessary. In his lectures on the treatment of injuries, he even explicitly encouraged a spirited approach. Some surgeons, he warned his students, wanted to spare their patients agonizing pain out of pity ("pietas"). But with dangerous injuries, one had to intervene and, if necessary, cut and amputate. This was still better than letting the patient die.[91] Yet, one of the most striking features of Falloppia's teaching on the treatment of ulcers and tumors was the caution and restraint he often recommended. The good surgeon, he made his students understand, knew not only how to perform surgical procedures. He also knew when it was better to refrain from them because of the nature of the condition, the state of the patient, and the risks involved.[92]

Falloppia urged particular restraint in the case of ailments whose treatment was known to fail frequently. Paradigmatic for this were cases of ulcerating cancer.[93] In Falloppia's days, they were primarily diagnosed in women, because the diagnosis of cancerous diseases inside the body, as they also occurred in men, was rarely possible during their lifetime. Especially breast cancer was quite common and much feared, because of the pain, the putrefying flesh, the stench, and the ultimately often fatal outcome.[94] Falloppia outlined the various therapeutic options. Some remedies were for local treatment, others served to divert the flow of cancerous matter that maintained the ulcer and to evacuate it through other pathways. When he presented the case of a distinguished lady with a rapidly growing hard, cancerous tumor in the right breast in a *collegium*, in 1552, he opposed a surgical intervention and recommended the administration of *purgantia*, bloodletting, and local remedies instead.[95] Other remedies were chosen to counteract the formation of cancerous matter at its source. In this sense, Falloppia praised, for example, the beneficial effects of a decoction of cinchona root. Only last winter he had given it to a distinguished patient in Venice who was suffering from a very painful cancerous tumor of the breast. Already after eight days, the pain had disappeared.[96]

If a surgical removal was indicated, Falloppia explained, the patient had to be bedded in a suitable manner, also to provide for the event that she fainted, and she was to be given wine. When he cut the cancer out with a knife, the surgeon had to make sure he extirpated the cancer with its roots. Ulcerating cancer usually took its origin from glandular tubercles which hardened and ulcerated over time. It was therefore not enough to remove the flesh around the visibly ulcerating parts. The surgeon also had to explore the deeper regions of the breast to see if he could find other ulcerating glandules or at least glandules that already seemed swollen and affected by the cancer and remove them as well. If he neglected to do so, the remaining glandules would develop into cancer again, within a few months, and break through the skin to the outside. After the operation, the surgeon had to stop the bleeding. Recently some surgeons had started using cautery for this purpose but Falloppia gave preference to (less painful) astringent agents whenever possible.[97]

Falloppia described the careful removal of the cancerous tumor together with the cancerous glandules in the surrounding area in considerable detail but he added that he himself, like many other physicians, avoided the use of knives and fire as much as possible in such cases.[98] He preferred using of corrosive substances, which he applied to the cancerous tumor or, if that was possible, to separate the cancerous tumor, which had been lifted from the surrounding flesh with a thread, from the remaining tissue.

After this presentation of different therapeutic options, Falloppia's lecture took a remarkable turn. It is expressed, albeit in somewhat different terms, in Bruno Seidel's edition of his lecture notes of 1577, in the edition of *De ulceribus* in the collected works of 1606, which was probably based on Marcolini's notes, as well as in the manuscript notes an unidentified student took the last time Falloppia lectured on this topic, in 1660–1661. Out of a hundred patients with ulcerating cancer, Falloppia told his students, hardly one recovered.[99] Even when the cancer was incurable, the patient must not be left to his fate, however.[100] In these cases, the physician or surgeon had to resort to a "cura" or "curatio" "paleativa", "palleativa" or "palearis" – Falloppia used various spellings. It did not aim at healing but at slowing down the progression of the disease, fighting the symptoms and relieving the pain.[101] The notion of a "palliative" treatment is widely perceived as a recent, modern creation but it can be traced back to the late Middle Ages, to the work of Guy de Chauliac. In the course of the sixteenth century, terms like "cura palleativa" and to "palliate" gradually gained currency.[102] Falloppia's surgical teaching and the publication of the respective lecture notes in Latin and vernacular may well have played a major role in this.

The "palliative" treatment of cancer sores, Falloppia explained, demanded among other things, to dry out the ulcer, to drain the morbid secretions, and to strengthen the affected limb so that it would absorb the morbid matter less easily. The means by which this could be effected were plain and mild. Goat's, sheep's, or cow's whey, or real milk into which red-hot steel had been repeatedly dipped for a short time, were very suitable for this purpose, as was the juice of finely ground nightshade (*Solanum hortense*).[103]

Pain relief deserved special attention in patients with ulcerating breast cancer. Some of them constantly "plagued" ("excrucient") the doctors with requests for new medicines against the pain. The same was true for women who had painful cancerous growths in their genitals. The surgeon therefore had to have a certain arsenal of painkillers at his disposal. If various remedies had already failed, he could secure the great gratitude of the women if he finally did find a remedy that dulled the pain. They would virtually worship him. As the lecture progressed, Falloppia listed a number of remedies that could relieve the pain, when local treatment was insufficient, including opium and hyoscyamus.

Sometimes, unfortunately, nothing could be achieved even with powerful opiates. Then one could only ask God to let the sick person die. Because it was still better to die once than to have to endure great pain for such a long time.[104] "So if you can palliate, then palliate" was Falloppia's conclusion according to the Venetian edition of his lectures, "if not, then pray for the death of the patient."[105]

Falloppia's detailed discussion of palliative care seems to have impressed the students. Bruno Seidel was later one of the first authors to devote a separate monograph to the treatment of incurable diseases.[106] He strongly criticized the imprudence and vanity of not a few – and sometimes very experienced – physicians who boasted of their successful treatment of a fully developed cancer. He did not believe that anyone had ever been cured of such a well-developed, ulcerating cancer. When such external cancers were cut out, they only came back – or even became much worse than before.[107]

Cosmetic medicine

One of Falloppia's most original surgical lectures was devoted to "medical cosmetics" in a broader sense. In 1566, four years after Falloppia's death, Petrus Angelus Agathus published his notes on them under the title *De decoratione* in his edition of Falloppia's *Opuscula*.[108] After some preliminary considerations on different definitions of beauty and on the question to what extent the preservation and, if applicable, restoration of beauty was the task of medicine at all, Falloppia discussed the treatment of obesity and its opposite, emaciation ("macies"). In the following chapters, he turned to the treatment of a short foreskin, the lengthening of a short penis, the treatment of small and excessively large testicles, the surgical treatment of mutilated or malformed noses, and the treatment of breasts and nipples that were unattractive and/or could make breastfeeding difficult.[109] The notes end with Falloppia's announcement that he would eventually also deal with changes of the skin color of the body as a whole and of the individual parts of the body, as well as with discolored spots that "infected" the skin ("inficientibus cutim"), with head lice, and other important topics. As Agathus explained in his epilogue to the reader, Falloppia's untimely death prevented him from bringing this plan to an end. Agathus' remark thus suggests that Falloppia delivered this lecture in the academic year 1561/62 and intended to continue it in the following academic year.

With his lecture, Falloppia entered new territory.[110] It is the first independent scientific treatise by an academic physician on this subject. Falloppia's treatise was undoubtedly a source of inspiration for the treatise *De decoratione* by Girolamo Mercuriale, a work that was also published based on a set of student notes, in this case, those of Giulio Mancini. Mercuriale explicitly referred to Falloppia's work in his lecture.[111]

Falloppia began by distinguishing between the good arts, which strove for the perfection of the substance or essence of things, and the bad arts, which merely aimed at an appearance of good and perfect. The goal of medicine was the perfection of its object, the human body. There were different views on what this perfection consisted of but these views could be reconciled. Some saw the perfection of the body in its ability to perform the bodily functions. Insofar as the body was, with Galen, the instrument of the soul, the highest perfection of the body was indeed its strength. Others regarded health as the highest good. Others again declared beauty to be the highest physical good, which not only women but also men strove for, young and old.

Ultimately, Falloppia concluded, despite this apparent diversity of views, physical perfection was a single good, composed of health, a good habitus, beauty, and the ability to perform the necessary functions. The different definitions put only different aspects in the foreground. Medicine served all these goals and thus surpassed other arts, such as gymnastics, athletics, and makeup. Since beauty was part of this more general good of physical perfection, it followed that the preservation and restoration of beauty was the task of medicine ("pulchritudinem pertinere ad artem medicam").

Some philosophers, Falloppia acknowledged, had passed a negative judgment on beauty, arguing that its effects were bad rather than good. Drawing on Galen's Thrasybulos,[112] Falloppia, like Mercuriale after him, introduced a fundamental distinction, between genuine natural beauty and false, preternatural beauty. The latter, a fictitious, whorish beauty produced by makeup and the like, destroyed what was naturally given and put something artificial and evil in its place. The pursuit of this kind of false beauty made women destroy their natural beauty, deprived old people of their dignity, and made young men effeminate.[113]

It is striking, also in view of modern debates, that Falloppia not only decidedly advocated "natural" beauty but also rejected the idea of an absolute ideal of beauty valid for all people in favor of an individualizing view. The beauty that medicine was to help preserve and, if necessary, restore was the beauty that corresponded to the person's age. Attempts to cover up one's age with makeup, Falloppia's made clear, created a fictitious beauty and had no place in legitimate *ars cosmetica*.

Falloppia expounded this ideal of natural beauty at length. Some conditions were "turpis" per se. The judgment on others conditions, by contrast, varied according to the beauty ideals that governed in the respective culture. The customs and with them the standards for the evaluation of beauty, Falloppia underlined, were not the same everywhere. In some places, women preferred small breasts because they seemed more modest or chaste ("modestiores"). In others, they desired plump, fat breasts. Exposition to the eyes of others also could play a role. Imperfect male genitalia were a much more powerful source of embarrassment and concern in antiquity, Falloppia found, because then men commonly showed themselves naked in baths or when they were wrestling.

Medicine could sometimes contribute indirectly, *per accidens*, to the preservation or restoration of beauty, by preventing or curing diseases that affected beauty. Medicine could also make beauty its primary goal, however. This part of medicine was called "medicina cosmetica", "ornatoria", or "decoratoria". In contrast to the *ars comptoria, medicina decoratoria* was exclusively concerned with natural beauty ("secundum naturam"), that is, with that beauty, which was natural to the human species, to a particular person with his or her particular habitus and age, or to this or that part of the body.

After these general considerations and a brief overview of the previous treatment of the subject in medical literature, Falloppia discussed a range of physical conditions and deficiencies where medicine could make a significant contribution.

Just like Mercuriale after him, Falloppia gave a prominent place to obesity.[114] Some people were obese by nature. In others, obesity did not result from their natural habitus but was acquired by habit. This obesity was not preternatural either, because, with Galen, habit ("consuetudo") was a second nature ("altera natura"). Truly pathological as opposed healthy obesity reached such an extreme degree that it damaged the functions ("actiones") of the body. Falloppia listed some of the harmful consequences that the medical literature had described since ancient times as a consequence of excessive obesity.[115] The extremely obese suffered from respiratory distress, heart tremors, and palpitations, and often also from diarrhea or dysentery. They risked dying prematurely. Since their constricted veins allowed only for a small volume of blood, they could not endure prolonged fasting. At the same time, their appetite was reduced. Their senses were weakened, and their mobility was severely restricted. Some could no longer reach their own posterior with their hands to clean it. Others were hardly able to touch their head. The extremely obese felt no sexual desire and did not enjoy sexual intercourse. With some notable exceptions, they were often infertile and when an extremely obese woman conceived after all, she easily lost her child or gave birth to a weak child. Fighting this morbid form of obesity was the task of medicine.

There were also obese and even very obese people, who were in good health. Their physical functions ("actiones") were at most weakened but not harmed. They could perform their tasks and duties, ride, walk, and travel. They had children, studied, and did all those things that a person living in society ("homo civilis") had to do. This "healthy" obesity thus did not damage the functions, which according to Galen was the decisive criterion for the definition of a disease. It did carry dangers for health in old age, however, and it was ugly or unseemly ("indecora") and indecent ("indecens") in the young. Therefore, the physician had to combat this "healthy" variety of obesity as well, which was difficult, however, and all the more so, the longer the obesity had already existed.

In order to treat obesity successfully, the physician had to know its causes. Insofar as obesity resulted from an excess of fat, the central cause was a cold and moist habitus. The production of blood in the body also brought forth a large quantity of fine, light, oily matter, which served to nourish the *calidum innatum*. Warm and dry bodies largely consumed this oily matter and the rest evaporated. In colder and moister bodies, by contrast, it permeated through the walls of the veins into the surrounding parts. In the warm, fleshy parts, it could ultimately evaporate but when the limbs were idle and their warmth was weak, the oily matter solidified and obesity arose.

The treatment of obesity therefore had to aim at bringing about a warmer drier habitus, at liquefying and draining the excess oily fluid, and at preventing the formation and accumulation of new fat. The cold habitus could be influenced by exposing the body to the sun, naked, if possible. In Falloppia's experience, the recommendation of Oribasius and Rhazes to cover the body with ashes or sand had also proved effective in natural obesity. Mud baths were likewise useful. Vigorous, sweat-promoting physical exercise, such as wrestling, horseback riding, and ball games, was particularly efficacious. Unfortunately, it came with the risk of falls and injuries. With Galen, walking was therefore recommended above all. Moreover, the obese should sleep rather little and on a hard bed or board and not in feather beds. Sexual intercourse was also very beneficial but the obese often had no desire for it.

The food intake should be sparse and ideally limited to one meal a day. The challenge was that some obese people were plagued by hunger. Falloppia had seen obese people who did everything to become thinner but who then ate like five men and devoured at least two capons at a single midday meal. In addition to eating less, the obese had to choose foods that were not very nutritious. Lettuce, spinach, onions, and the like could be eaten without any restraint. Water with vinegar was to serve as a drink. Falloppia did not share the reservations of other authors. He had seen obese people who had successfully drunk a jug of six or eight ounces of vinegar daily for forty days. Moreover, to promote the rapid evacuation of ingested food, enemas and laxatives were helpful.

While the body size of the pathologically obese was excessive, people on the other end of the spectrum were too thin. Three degrees of "macies" could be distinguished. In the mildest form, only the fat dissolved, in the second degree, the fat and the flesh were affected, and in the highest degree, the natural moisture was consumed as well. The major causes were those very things that were used to fight obesity: exposure to the sun, strong exercise such as swimming in the sea, running, jumping and ball games, and lack of sleep. Sexual intercourse was particularly harmful for the excessively thin. In some people, the *macies* resulted from diarrhea or from an excessive evacuation of bodily matter with the urine. Thus, the *diabetici* were characterized by *macies*. Among the important internal causes were an insufficient concoction of the food and, in children, worms. Even plump infants visibly emaciated because of worms.

The treatment of macies rested on a suitable regimen and warming from the outside. If the *macies* was not the pathological consequence of consumption and the physical functions were not disturbed, treatment was often unnecessary. Apart from some women who preferred to look obese and plump, most people liked to be thin anyway, because of better mobility.[116]

Sometimes only a certain part of the body became thin and emaciated. This was seen as a temporary phenomenon after injuries, for example, when someone broke his leg and had to keep still for forty days. Other causes included constricted vessels, which no longer let a sufficient amount of blood and vital spirits reach a certain limb. In these cases, treatment aimed at strengthening the part and, if appropriate, at seeking to direct the flow of blood toward it.[117]

In the remainder of his lecture, Falloppia turned to the enlargement of individual body parts and, in particular, to the issue of excessively large breasts. In some women, he explained, the breasts grew to such a size that they became ugly and embarrassing. Especially in unmarried women, this could have harmful consequences and require medical attention. Excessively large breasts could raise doubts about their virginity, because the size was attributed to their heating from carnal desire.[118] Even in women who already had children, small, round breasts were more beautiful than large, fat ones, however, which were reminiscent of the teats of a cow's udder.

The treatment of unsightly large breasts was primarily directed at the external causes. Among these, Falloppia counted, in particular, long baths, loose clothing, frequent sexual intercourse, and the touch of male hands. But also the abundant consumption of strong wine or of *zampiglione* – something resembling eggnog made from sweet wine, sugar, and egg yolks, among other things – could make the breasts heavier.

If this was not enough, the physician could resort to external remedies, especially astringent, cooling, and drying *emplastra*, which were changed every three days only so as not to stimulate the breasts and the flow of blood toward them by touch. Falloppia also mentioned a wide range of remedies, such as hyoscyamus, cicuta, and opium, that could be used. All of these remedies were surpassed, however, in Falloppia's experience, by crushed snail shells mixed with honey and the green roots of delphinium. Some authors also recommended the external application of blood from pig testicles or from hedgehogs or turtles but Falloppia preferred his "proven" remedies. He did mention one more, particularly powerful remedy, however, which an old woman had taught him ("vetula me docuit"). One had to apply the menstrual blood of a virgin two or three times on the breasts and leave them uncovered afterward. According to the woman, it had to be the blood from the very first menstrual period of a woman but in Falloppia's experience ("ego expertus sum"), it could also come from later ones.

Drawing on Paul of Aegina, Falloppia also described the surgical treatment of enlarged breasts with the knife, in women as well as in men, when the latter grew breasts. One had to make an arch-shaped ("lunar") incision at the top or bottom of the breasts, and in case of very large breasts, two intersecting incisions. From there, moving toward the nipple, the surgeon had to detach the skin and remove the underlying fat. If severe bleeding prevented this, a cauter could be applied. Wound healing was then promoted with an adhesive ("glutinatorio") agent. Falloppia described the procedure but added that he had no sympathy for such an agonizing procedure, no matter what the Greeks and Arabs thought of it. Even having breasts as large as wine jugs was better than that.

On the other end of the spectrum were breasts that were too small. They not only affected the woman's beauty but could also require treatment for health reasons, Falloppia explained, because the woman could not give her children enough milk. In that case, exactly the things that women with over-size breasts had to avoid were advisable: loose clothing, frequent and long baths, soft touching with the hand, and warming external agents such as mustard plasters.

Nipples that were too small were likewise both ugly and potentially harmful to health, Falloppia declared. They made sucking difficult or impossible for the infant and the milk thickened in the breasts and caused illness. Warming remedies, such as ointments, baths, and pulling the nipple with the hand, helped against this. The latter, Falloppia assured his students, did not damage the woman's decency ("honestate"), not even in unmarried women, but one could also use appropriate tools. Some women had invented an excellent device with which they could suck the milk from their own breasts when they did not have an infant that could relieve them. It was a glass vessel, similar to a cupping glass, but with a long beak through which the milk could be sucked. The same vessel could be used to enlarge the nipple. Others also used a reed, which they hollowed and cleaned in order to put it on the nipple and suck from the other side. Falloppia himself made use of a special tube made of lead, provided the breast had no ulcers or cracks and the woman had no milk in it. It was just wide enough on one side to fit over the nipple. On the other side, there was a small hole through which he sucked, pulling the nipple inside the tube. Then he closed the hole, first with tongue and hand and finally with wax, so that the nipple was held in the tube. It was important, he added, that the tube was not heavy and had no sharp edges.

Falloppia devoted further sections to the defects or alterations of the male genitals. Excessively large testicles made adolescents ugly, salacious, and unsuitable for singing.[119] Falloppia usually refrained from treatment, however. Surgical treatment with the knife was out of the question. If anything, conservative treatment, by tying the testicles to the body and immobilizing them, could be considered.[120]

Falloppia paid more attention to the treatment of a penis that was too small. One could blame him, Falloppia said, for talking about such an "obscene" topic. A sufficiently large penis was not just a matter of beauty, however. It was necessary for the preservation of the species. If the penis was not sufficiently large, the woman felt no pleasure, did not secrete semen, and did not conceive. Parents should therefore ensure from infancy that their son's genitals developed sufficiently. A frenulum that was too short could be in the way. Essential for sufficient size was, above all, frequent expansion. Coitus could contribute to this but was ultimately less suitable because it lasted only a short time. It was more effective to continuously stimulate the inflow of the *spiritus* responsible for the expansion. Arabic authors recommended daily rubbing with sheep's milk or even fats, especially fats like castoreum, which stimulated desire at the same time. Otherwise, the external application of warming agents and, if necessary, remedies that softened the tense nerves on the back of the penis helped.

Only rarely, two or three times, Falloppia, according to his own account, had also treated a shortened or defective foreskin. People no longer showed themselves naked in front of strangers who could have seen them. A lack of foreskin could even have advantages. Circumcised men rarely contracted the French disease because the skin of their glans became thick and hard, making it difficult for the "virulentia" to penetrate. An intact foreskin provided the lubricity, however, that was important for pleasure during intercourse, and when the woman felt greater pleasure she discharged her semen and the matter for the formation of the fetus and the membranes. Falloppia described various procedures for pulling the foreskin forward and holding it in this position. He also mentioned the method described by Paul of Aegina of lengthening the foreskin by making incisions in it and then fixing it so that it could be more easily pulled forward. Falloppia, however, had never done this himself, nor did he know anyone who would have been foolish enough to submit himself to this ordeal.[121]

In *De decoratione*, Falloppia expressed doubts about the use of the knife that seem remarkable for a surgeon. His skepticism also extended to what is probably the most famous surgical procedure for cosmetic reasons from that period, the reconstruction of the nose. Damaged, deformed, or outright missing noses were fairly common at the time due to war injuries and the French disease. They were more than just an esthetic problem. Cutting off the nose was a common type of punishment for criminals. A nose that was mutilated by injury or disease could thus also jeopardize the victim's honor.

In *De decoratione* and also in his lectures on traumatology, Falloppia described how some surgeons from Calabria proceeded in the surgical reconstruction of nose. Three decades later Tagliacozzi was to make the procedure famous.[122] Falloppia described it in detail. The surgeons superficially opened the skin of the nasal stump and on an upper arm. The upper arm was then tied tightly to the nasal stump and held in this position for several weeks until the arm and the nasal stump grew together. Eventually, the upper arm could be cut loose, leaving a flap of skin with vessels attached to the nose. From the flesh that came from the arm, something in the form of a nose was made. Falloppia was very skeptical, however. The patients, he pointed out, suffered months of agony. It was better to use an artificial nose than to endure such a torment.[123] All the more so, since the prospects of success were rather limited, as the students could see with their own eyes in the case of a goldsmith in Padua. The operation had given him a *nasellus* rather than a real nose.[124]

Trauma: wounds, dislocations, and fractures

Falloppia's traumatology has come down to us in different sets of notes lectures. In 1566, Luca Bertelli brought out Agathus' notes on Falloppia's lectures on the Hippocratic *On head injuries*.[125] Under the title *Opuscula tria*, Paulus and Antonius Meietus in Venice soon after published student notes on Falloppia's lectures on wounds in general and on injuries of the eyes and of other parts of the body, in addition to his commentary on the Hippocratic *On head injuries*.[126] In 1571, they printed Falloppia's traumatological lectures under the title *De parte medicinae, quae chyrurgia nuncupatur* or as it was called more briefly inside the book, *Libellus de vulneribus*, now adding Falloppia's discussion of gunshot wounds.[127] In the following, I will draw primarily on the particularly extensive notes, which the brothers De Franciscis published in 1606, in the second volume of their edition of Falloppia's *Opera*.[128]

As in his surgical teaching in general, these lectures were very much oriented toward practice and peppered with numerous prescriptions. At the same time, drawing on his training in the liberal arts and his command of Aristotelian philosophy, Falloppia like other learned physicians who turned toward surgery, placed great emphasis on a rational, methodical approach. This was the decisive basis for the claim to practice surgery like medicine as a whole as "scientia" and not only as an empirical art.[129]

In his lectures on the wounds, Falloppia embarked on a particularly elaborate theoretical discussion. He began with the common definition of injuries as a "solutio unitatis" or "continuitatis" and investigated the question of whether this "solutio continui" actually affected the *partes similares* – roughly speaking, in modern terminology, the different kinds of tissues – and the organs or *partes instrumentales* in equal measure, as the Galenic orthodoxy claimed. Some of his Paduan colleagues tried to refute his opinions with "futile arguments" – we will come back to this – but his conclusion was clear: a *solutio continui* was a disease of the *partes instrumentales*, whose unity was lost and whose function ("operatio") was disturbed or completely lost.[130]

In a second step, he differentiated different types of injuries like *contusio*, *fissura*, *perforatio*, and *scissura*. A special case was *anastomosis*, here understood as the excessive widening and opening of the small endings at the extreme end of tiny vessels or *capillamenta*, as it occurred in severe bleeding from the nose and the hemorrhoidal veins.[131] Finally, wounds had to be distinguished and classified according to their causes, location, type, size, appearance, and effects.[132]

Falloppia then dealt with the prognosis and treatment of wounds in general. Some wounds, for example, cuts in the longitudinal direction of muscles, healed easily, "merely with spit", as people said.[133] Others were usually lethal. Even minor injuries to the heart, for example, were almost never survived. Injuries to the brain, by contrast, were dangerous but did not always lead to death. Hippocrates had used misleading words here.[134] Already Galen and Guy de Chauliac had reported cases of patients who survived such injuries. Berengario da Carpi claimed that he alone had healed six men with head injuries in which parts of the brain had been cut off.

Regarding the treatment of wounds, Falloppia described in detail various types of dressing and bandaging and gave his students recipes for remedies that would promote the healing of the wound; mastic, plaster, gummi arabicum, and other substances could be used here.[135] He himself preferred a very simple remedy that was easy to prepare: he mixed some finely ground lime ("calcem vivum") with beaten egg-white and applied the mixture with a piece of cloth.[136]

To suture wounds, it was best to use linen or silk thread. Thread made of cotton or wool rotted too quickly. Falloppia also advised stitching where others often dispensed with it. For example, it was controversial whether or not to suture lacerations of the scalp. Some simply cut off the detached skin because they thought it would not heal. But, according to Falloppia, "daily experience" showed that it was wrong to refrain from suturing head wounds.

With larger wounds, one had to add *sarcotica*, medicines that promoted the growth of flesh and thus helped to fill the defect created. And one had to try to cover this defect with skin afterward.

When a foreign body had penetrated into a part of the body, the surgeon could seek to either extract it with suitable instruments or give medicines that would attract it and pull it to the outside by a "proprietas occulta", which resulted from their "total substance". Or he could use remedies that caused heat or putrefaction and thus promoted the elimination of the foreign body.[137]

It was not always necessary or indicated to remove foreign bodies, especially when their removal posed considerable risks. Albucasis had reported of men, for example, who lived for years with an arrow or an arrowhead in their body. Falloppia himself had removed the lead bullet from the body of a man who had been shot five years before. Because of the heavy bleeding, the physicians had not been able to remove the bullet at the time. Meanwhile, the lead had moved from the original place of the wound, in the groin, to the middle of the thigh. In another case, it was nine years after the injury, that he pulled out a piece of iron from the thorax of a patient. He mentioned other cases of this kind.[138]

After an overview of the various complications of wounds, such as inflammations, swelling, and gangrene, in general, Falloppia moved on to the injuries of the different parts of the body.

Head injuries deserved special attention here. A dangerous peculiarity of head injuries was that they were sometimes fatal although no major injuries were visible from the outside. The skin might even remain intact but the brain inside, the seat of the *anima sensitiva* and *rationalis*, was injured.[139] Only the year before, Falloppia recounted, a young Roman had died after watching others play ball. When the big ball came at him, he tried to avoid it and fell on his head. He was dazed, could no longer speak, and died four days later.[140]

Moreover, the brain sometimes suffered damage not only on the side of the impact but also on the opposite side; this corresponds to modern medical doctrine of the so-called contrecoup. As Falloppia explained, the brain did not completely fill the cranial cavity most of the time, except with the full moon, when the brain swelled, as he himself had observed in his dissections. As a result, some space remained on the side opposite the traumatic impact, and the brain was thrown against the bony wall there by the impetus and was harmed.

For the diagnosis of skull fractures, the surgeon could use a probe. When a head wound penetrated the scalp, he could also ask the patient to hold his nose and breathe out through the nose against the pressure. If there was a skull fracture, some blood or fluid could be seen leaking out of the hidden fissure. Following the advice of Hippocrates, one could also apply a cloth dyed with ink or the like. If there was a fracture, it would be recognized the following day by black edges.[141]

When brain matter emerged from the fracture in open skull injuries, the fungus-like excrescence that developed should not simply be cut off, as some surgeons did. Falloppia advised surrounding the area with a protective covering, instead. It was quite possible that the extruding matter would return into the inside of the skull.

Even open skull injuries were not always immediately fatal. Some patients lived for three weeks or even sixty or ninety days. Falloppia knew of two patients who were still alive after such an injury.[142] Only recently he and other surgeons had treated a young man in Padua who lost brain matter the size of a walnut through a head injury and was cured.

Perforating eye injuries almost always resulted in blindness.[143] Foreign bodies which had penetrated into the eye interior and could not be easily be removed with a forceps. It was better to leave them there, if they were not poisonous or rusty. Falloppia retold one of his favorite stories in this context: a woman had a glass jar full of feces thrown in her face. A small shard entered the eye and Falloppia could not safely remove it but in the end the woman could see with the injured eye, but everything seemed twice as big, probably because the light in the piece of glass was "doubled" ("reduplicatur"), he thought.[144]

In the case of nose injuries, the surgeon was not to give up too quickly. Even when the nose was only hanging on a scrap of skin, it was worth a try, because the nose tended to heal well. Falloppia felt that the nose could probably even heal if it had been completely severed and fallen to the ground. It just had to be put back into its place very quickly, before the blood and spirit disappeared from the severed part. Falloppia even toyed with the idea of a heterologous transplantation of the nose, as we would call it today. If someone's nose was cut off and one quickly attached the cut-off nose of another person before the spiritus disappeared from it, then this nose would grow in and the patient would have a new nose. Falloppia did not elaborate on where this other nose should come from. Perhaps he was thinking of people who had their noses cut off as punishment or had just been executed.[145] Falloppia also described again the surgical reconstruction of the nose by means of a flap of flesh and skin from the upper arm, expressing his reservations about this excruciating procedure.[146]

The discussion of injuries to the neck and the neck vessels offered Falloppia an opportunity for a more general discussion of the treatment of severe hemorrhage. Hemorrhage, even severe hemorrhage, was not fundamentally harmful, he explained. Sometimes, in the case of a surplus of blood, it even had a beneficial, liberating effect, as his audience could see in the case of the theologian and Dominican friar Jerome Monachus, who ejected more than three pounds of blood from his chest and recovered.[147] Bleeding from larger vessels, however, was very dangerous. It often could not be stopped by astringents and other medicines that were commonly used for nontraumatic bleeding. Though this was not common practice, it was best in such cases, if possible, to tie the injured limb above the wound with a tight bandage in order to diminish the afflux of blood. Then the vessel could be laid open and isolated from the surrounding flesh and from the nerve that regularly accompanied it so that one could wrap a thread around the vessel and stop the bleeding.[148]

In the case of thoracic and abdominal injuries, it was necessary to check whether the wound extended into the interior of the chest. In that case, the organs and structures there might be injured, too. When abdominal injuries were not accompanied by injuries to the liver, stomach, intestines, or bladder, Falloppia claimed, he had always been able to heal them. When parts of the caul (*omentum*) protruded through the wound, bystanders and inept surgeons sometimes did not dare push it back into the abdominal cavity but this was the thing to do. Once, Falloppia reported, he had even removed a man's entire omentum without adverse consequences. This was surprising in that the caul, according to Galen, maintained the vital heat, the *calor naturalis*, but the man could preserve his vital heat without omentum.[149]

Stomach injuries were to be sutured only if the wound was easily accessible. Intestinal injuries, on the other hand, always had to be sutured, even though, as Guido Guidi described in one case, the wound could sometimes miraculously close by itself. Falloppia's students would still remember that woman with an abdominal wound from which the feces exited. Because only her abdominal wall was sutured but not the intestinal wound, the feces accumulated in her abdomen and eventually emptied through other routes.[150]

Falloppia extensively discussed different techniques of treating the intestinal wounds. Some he did not commend. Certain military surgeons, for example, who often encountered intestinal injuries in the battlefields always carried dried pieces of animal intestine with them to close the injury. Others used the trachea of geese or swans for this purpose.[151] Falloppia thought nothing of this. Dried intestines would soon contract and rot. He advised a thorough cleansing the intestine followed by an ordinary suture, with separate stitches at a small distance from each other.[152]

A major new challenge were injuries caused by the musket bullets and cannon balls. One wished, Falloppia exclaimed, that they had never been invented and disappeared from the earth again. Cannon shots ("bombardae") were mostly fatal. Only sometimes, when the injury was limited to an extremity, treatment by way of amputation was indicated.[153] With gunshots, by contrast, the lead bullets could be removed with special instruments. He praised in particular the so-called alphonsinum, which the Italian military surgeon Alfonso Ferri (1515–1595) had invented for this purpose. It was inserted along the trajectory of the shot until it reached the bullet. At that point, the surgeons opened two or three claws by moving a ring on the handle and sought to grasp the bullet and extract it. As Falloppia freely admitted, however, he had not yet used this "excellent instrument" himself.[154]

He rejected the idea that the injuries caused by firearms were particularly dangerous because the bullets were very hot or because the gunpowder that still adhered to them was poisonous. Bullets that were fired into wax did not melt it. They were not hot. And that the gunpowder was not poisonous was clear from the accidents that frequently occurred when people lit fireworks and powder was hurled into their faces, penetrating the skin and sticking there. Only recently, Falloppia had been consulted about a boy in Venice who had made a firecracker out of gunpowder and paper. It had exploded prematurely and numerous grains of powder had entered his cheeks, remaining there for three years without doing any discernible damage to his health.[155]

To the injuries of bones and joints, and, in particular, to fractures and dislocations, Falloppia devoted another separate series of lectures. Students notes were later published as *De luxatis et fractis ossibus* or separately as *De luxationibus* and *De fracturis*. Compared to the lecture on wounds, he drew much more on the extant literature in this field than on his own experience. Even with regard to practical details, he quoted Hippocrates, Galen, Celsus, Paul of Aegina, and other ancient authors. Exceptionally only he mentioned actual cases from his own experience, such as that of the Canon Barisono, who just recently had suffered a dislocation of the right upper arm from a fall,[156] or that of the son of the Venetian patrician Stefano Tiepolo, whose broken leg he had set; the healing progressed well at first but then his condition worsened in the seventh week due to the patient's grief over the death of his father, as Falloppia thought.[157] As in all his lectures, he provided his students with various recipes but his therapeutic recommendations in this area were only rarely explicitly based on his own experience.[158]

There may be a reason, why Falloppia referred much less to his own experience, when it came to fractures and dislocations. His services were probably not often demanded in such cases. In Italy, as in other regions of Europe, there were specialist bonesetters, who knew how to set and splint broken bones in humans and animals. According to Falloppia, these "nostri restauratores" deserved to be called "corruptores", because they splinted fractures immediately, instead of waiting until the seventh, ninth, or eleventh day.[159] Falloppia must have been aware, however, that some of these practitioners enjoyed the confidence of the population and often had acquired great experience in dealing with fractures in animals. Some executioners also had considerable knowledge and experience in this field.[160] In his treatise on shoulder dislocations, Falloppia was critical of them. Dislocated shoulders, he pointed out, were frequent among people who were submitted to torture, when the torturer tied their hands behind their backs and pulled them up with a rope to their hands in order to get the truth out of them. If these men knew more about the anatomy of bones and the art of repositioning dislocations, not so many people would be permanently deprived of the use of their arms after the ordeal.[161] Interestingly, he did converse with these people, however. They "taught" him ("docuerunt me"), he reported, that it was best to use special strings made of goat's leather, as it was used for clothing, the so-called coletto, to tie up the limbs that had to be stretched for repositioning. Those made of fabric tore and the leather of calves and oxen was too hard.[162]

Notes

1 Daza Chacón, Pratica (1673), book 2, p. 232.
2 Murhardsche Bibliothek, Kassel, 4 Ms. med. 19, foll. 1r–140r, students notes on Falloppia's lectures on ulcers and the French disease, 1555; Wellcome Library, London, Western Manuscripts 269, student notes on Falloppia's lectures on ulcers and the French disease, 1560–1561; Bodleian Library, Oxford, MS. Canon. Misc. 115, student notes on Falloppia's lectures *De luxationibus* and *De fracturis*; Российская Национальная Библиотека (Russian National Library), St. Petersburg, Lat. F VI 95, with student notes on Falloppia's lectures on tumors; Biblioteca comunale, Urbania, Ms. 95, student notes on Falloppia's lectures on the French disease (incomplete), ulcers of the eye, and injuries, 1561; ÖNBW, Cod. 11225, foll. 25r–62r, Handsch's notes on Falloppia's lectures on tumors 1552–1553; Herzog-August-Bibliothek, Wolfenbüttel, Cod. Guelf. 22 Aug. 4°, includes student notes on lectures by Falloppia on cauters, tumors, ulcers, the French disease, and injuries, 1555–1556, and ibid., Extravagantes 264.2, foll. 10r–125v, student notes on Falloppia's lectures on wounds. The bibliographical information on the manuscript in St. Petersburg is from Kristeller's *Iter italicum*; I have not yet seen the manuscript myself.
3 La chirurgia di Gabriel Falloppio modonese [sic]. Venice: apresso Daniel Zanetti 1602 (but with a preface by the printer dated 1603); La chirurgia di Gabriel Falloppio modonese. Venice: Presso Giacomo Antonio Somascho 1603.
4 Falloppia, La chirurgia (1647); Falloppia, La chirurgia (1675) (three editions in the same year), one by Steffano Curti, one by Paolo Baglioni, and one by Abbondio Menafoglio (I have not seen the Menafoglio edition, which is listed in various library catalogues, however, e.g. in the Biblioteca nazionale in Turin).
5 British Library, London, Ms. Sloane 3293, English translation of Gabrielle Falloppia's "De ulceribus" and "De tumoribus praeter naturam" by James Molins.
6 McVaugh, Rational surgery (2006).
7 Chauliac, Chirurgia magna (1997).
8 McVaugh, When universities (2016).
9 Kinzelbach, Erudite and honoured artisans? (2014).
10 Gurlt, Geschichte der Chirurgie (1898), pp. 361–404.
11 Falloppia, Libelli duo (1563), fol. 5v: "Moneo ut chirurgus non tantum chirurgus esse velit, sed integerrimus medicus."
12 On Guidi see Preti, Guidi (2004).
13 Tagault, De chirurgica institutione (1543); in 1544, Valgrisi published a second edition in Venice, with a dedicatory epistle by the Padua professor Bassiano Landi. On Tagault see Biesbrouck, Goddeeris and Steeno, Jean Tagault (2017).

14 For an overview see Nutton, Humanist surgery (1985).
15 Cf. Stolberg, Bed-side teaching (2014).
16 Falloppia, De vulneribus (1606), p. 340.
17 ÖNBW, Cod. 11210, fol. 144r; SUBG, Ms. Meibom 20, fol. 158r.
18 ÖNBW, Cod. 11210, fol. 145r–v; SUBG, Ms. Meibom 20, fol. 157v.
19 SUBG, Ms. Meibom 20, fol. 156r.
20 Falloppia, Compendium (1585), fol. 16r.
21 Falloppia, De tumoribus (1606), fol. 1v.
22 Bertelli's dedicatory epistle is from April 1563. He published a second, revised edition in 1566.
23 Falloppia, Gabrielle: *De morbo gallico liber absolutissimus.* Ed. by Petrus Angelus Agathus. Padua: apud Lucam Bertellum & socios 1563.
24 Erfurt: Georgius Baumanus 1577.
25 Wellcome Library, London, Western Manuscripts, Ms 269.
26 Falloppia, Libelli duo (1563), foll. 1r–32v.
27 The text of *De ulceribus* in the 1606 edition of Falloppia's words is not identical with the one in Seidel's edition. The students may have attended Falloppia's lecture on this topic in different years. A rough comparison does not reveal any major differences, however, between the way he discussed the different ulcers and skin lesions and their treatment. The major difference in terms of content is that the 1606 edition has a few more chapters on specific subtypes of ulcerous lesions.
28 Falloppia, Libelli duo (1563), foll. 1r–3r.
29 Ibid., foll. 3r– 4v. For concrete evidence that some female patients at least were not too embarrassed or ashamed of the eyes and hands of a male practitioner, see also Stolberg, Learned physicians (2022), pp. 485–487.
30 Falloppia, Libelli duo (1563), foll. 6r–8r.
31 Ibid., foll.12r–19v.
32 Stolberg, Learned medicine (2022), pp. 265–268.
33 Falloppia, De ulceribus (1577), pp. 99–106.
34 Ibid., pp. 189–196.
35 Ibid., p. 235.
36 Falloppia, De cauteriis (1570).
37 Falloppia, De ulceribus (1577), p. 49.
38 Falloppia, De cauteriis (1570), fol. 72v.
39 Ibid., foll. 71v–72r.
40 Universitätsbibliothek Basel, Frey-Gryn Mscr. II 5, Nr. 9, "De cauterijs"; the Basel catalogue does not link the manuscript to Falloppia but there is hardly any doubt that it comes from student notes on his lectures. The text cannot be a mere excerpt of the printed text either: it is very similar to it but far from identical and on certain points more detailed.
41 Falloppia, De cauteriis (1570), foll. 70r–71r, "vt autem quae dicta sunt, ob oculos habeatis, facillimam vobis tabulam ac simplicem tradam"; the diagram can also be found, probably copied by Theodor Zwinger, under the title "De cauterijs" in Universitätsbibliothek Basel, Frey-Gryn Mscr. II 5, Nr. 9; on the uses of dichotomic tables and other diagrams in contemporary medicine in general, see Maclean, Diagrams (2006).
42 Arrizabalaga/Henderson/French, Great pox (1997).
43 According to Scott, Catalogue (1913), p. 85, Marsh's Library in Dublin owns an English translation of Falloppia's lectures "De morbo gallico" (Ms Z 4. 5. 2aa, 1576). I have not yet seen this manuscript myself.
44 Falloppia, *De morbo gallico liber absolutissimus* (1563). Some library catalogues give 1564 as the year of the first edition – and both dates are probably correct to some degree. The frontispiece as well as the cover page of Fracanzano's treatise are dated 1563 and Agathus' dedicatory letter is from November 1563 but at

the end of the book, we find "Patavii, apud Christophorum Gryphium M.D. LXIIII", which may indicate that the printing was finished in 1564 only. I have not found any evidence for a different 1564 edition. Bertelli soon after also published the work as *Tractatus de morbo gallico* in Falloppia's *Opuscula* (1566).

45 Falloppia, De morbo gallico (1565).
46 Falloppia, *De morbo gallico* (Sessa, 1574); idem, *De morbo gallico* (Regazola, 1574). Both editions only differ in the frontispiece, which names Sessa and Regazola respectively as the publisher. According to the colophone, the Sessa edition was also printed by Regazola.
47 Falloppia, Opera (1600), vol. 1, pp. 682–749.
48 Falloppia, Opera (1606), vol. 2, pp. 113–203.
49 Falloppia, De morbo gallico (1563), fol. 52r.
50 According to Seidel's edition of Falloppia, De ulceribus (1577), p. 1, the lecture started on 3 November 1557.
51 Wellcome Library, London, Western Manuscripts Ms. 269; the notes document only the beginning of Falloppia's discussion of the French disease: having obtained his doctoral degree, the (unidentified) student left Padua already in March of 1561.
52 Fracastoro, Syphilis (1536).
53 For a summary of Falloppia's understanding of the French disease (in the light of the medical knowledge around 1900) see Casoli, Sifilografi (1905), pp. 19–28.
54 Falloppia, De morbo gallico (1563) fol. 13r: "Morbus gallicus nouus est."
55 Ibid., fol. 2r.
56 Ibid., fol. 2v.
57 Ibid., foll. 5r–6r.
58 ASV, Riformatori allo studio, filza 63, draft of a letter to the rectors of the University, 7 August 1552, on an advance on his salary for Cadena who read "la lettion di astrologia".
59 Falloppia, De morbo gallico (1563), fol. 16v.
60 Ibid., fol. 8r and foll. 14v–16v.
61 See the chapter on pharmacology below for a more detailed discussion of the concept of "tota substantia".
62 Falloppia, De morbo gallico (1563), fol. 18r.
63 Ibid.
64 Ibid., fol. 19v.
65 Ibid., fol. 18v. Modern scholarship widely assumes that the clinical picture of the French disease evolved with time (Arrizabalaga, Henderson, and French, Great pox (1997), pp. 25–27).
66 Falloppia, De vulneribus (1606), p. 407.
67 Falloppia, De morbo gallico (1563), fol. 15v.
68 Ibid., fol. 17r.
69 Ibid., fol. 17r–v.
70 Ibid., fol. 30r: "Nam ad it, quod a tota substantia laedit requiritur medicamentum a tota substantia sanans."
71 Ibid., fol. 26v.
72 Ibid., fol. 42v.
73 Ibid., fol. 41v.
74 Falloppia, De morbo gallico (1563), foll. 52r–53v, ch. 89, "De praeservatione a carie gallica".
75 Himes, Medical history of contraception (1970), pp. 188–190; Youssef, History of the condom (1993), p. 226; Tsaraklis et alii, Preventing syphilis (2017); all three give 1564 as the date the first publication; see also https://en.wikipedia.org/wiki/Condom "Condoms as a method of preventing STIs have been used since at least 1564" (accessed 8 August 2021).

76 Agathus' notes were probably based on Falloppia's lectures in 1557/58. He referred to a lecture on the same topic, however, which Falloppia delivered in 1555 (i.e. presumably in the academic year 1554/55) in which he already recommended the use of this "linteolum" but gave a different recipe for its preparation.

77 Biblioteca comunale Urbania, Ms 95, foll. 5v–6v, student notes on Falloppia's lectures on the French disease, June 1561.

78 Since it has often been misinterpreted I quote the entire passage:

> Ego dixi quod nascitur caries haec per communicata corpuscula saniosa, quae imbibita poris glandis faciunt cariem, ideo opus est, ut statim saniem a glande expurgemus, sed si imbibita sit in poris, licet uino, lotio, uel aqua detergamus priapum, tamen eam detergere non possumus. & hoc saepe accidit in tectis, & mollibus glandibus. Quomodo ergo agendum? semper fui istius sententiae, quod ponamus aliquod habens uim penetrandi corium, & dissipandae materiae, uel extrahendae, uel siccandae et uincendae natura sua. ideo inuestigaui hoc medicamentum. Sed quia oportet etiam meretricum animos disponere, non licet nobiscum unguenta domo afferre. propterea ego inueni linteolum imbutum medicamento, quod potest commode asportari, cum faemoralia iam ita vasta feratis, ut totam apothecam uobiscum habere possitis. Quoties ergo quis coiuerit abluat (si potest) pudendum, uel panno detergat: postea habeat linteolum ad mensuram glandis praeparatum; demum cum coiuerit ponat supra glandem, & recurrat praeputium: si potest madere sputo, uel lotio bonum est, tamen non refert: si timetis, ne caries oriatur in medio canali, habeatis huius lintei inuolucrum, & in canali ponatis[…].

79 According to Biblioteca comunale Urbania, Ms 95, foll. 5v, he did not recommend urine but wine for this purpose in 1561.

80 More than a century ago, C. E. Helbig already arrived at this conclusion (Helbig, Zu dem Schrifttume (1907); idem, Geschichte (1913)).

81 Urbania Biblioteca comunale Ms 95, fol. 5v, "cum aliquis coiuerit cum muliere".

82 The form can be both the subjunctive of the present perfect and the future perfect. The preceding temporal conjunctions "cum" and "quoties" demand the indicative, however. So we are dealing with a future perfect.

83 Himes, Medical history of contraception (1970), p. 190

84 Falloppia, De morbo gallico (1563), foll. 52r, "ego feci experimentum in centum, & mille hominibus, & Deum testor immortalem nullum eorum infectum"; this claim is missing in the student notes from 1561 (Biblioteca comunale Urbania, Ms 95, foll. 5v–6v).

85 Falloppia, Libelli duo (1563), foll. 33r–101r.

86 Falloppia, De tumoribus praeter naturam (1606), foll. 1r–109v.

87 Ibid.; cf. Imperatori, Fallopius (1948), pp. 445–446.

88 Falloppia, De tumoribus praeter naturam (1606), foll. 38r–39v; Falloppia, Libelli duo (1563), foll. 63r–65v.

89 Falloppia, De tumoribus praeter naturam (1606), fol. 60r–v; see also Falloppia, Libelli duo (1563), foll. 81v–82v.

90 Falloppia, De tumoribus praeter naturam (1606), foll. 95r–96r.

91 Falloppia, De vulneribus (1606), p. 338.

92 Falloppia, Libelli duo (1563), fol. 5v.

93 Falloppia, De ulceribus (1577), pp. 91–106; Falloppia, Appendix (1606), pp. 17–27.

94 Stolberg, Learned physicians (2022), pp. 267–268.

95 Notes of Johannes Brünsterer, Universitätsbibliothek Erlangen, Ms. 910, foll. 50r–51r.

96 Falloppia, De ulceribus (1577), p. 99.

97 Ibid., p. 100.

98 Ibid., p. 101.
99 Ibid., p. 102.
100 Wellcome Library, London, Western manuscripts, Ms. 269, fol. 59v.
101 Ibid.; Falloppia (1577), p. 102.
102 Stolberg, "Cura palliativa" (2007); Stolberg, History of palliative care (2017).
103 Falloppia (1577), ca p. 102.
104 Falloppia, De ulceribus (1577), p. 106, "precandus est Deus vt morte liberet mise-
rum aegrotantem, melius enim longe est mori quam isto modo excruciari." Idem,
De ulceribus (1606), p. 77, "tunc rogandus est Deus Opt. Max. ut vita aegrum
privet"; "satius est semel mori, quam tantum perferre dolorem et tam longo
tempore."
105 Falloppia, De ulceribus (1606), p. 77: "Si potestis ergo palliare, pallietis: sin
minus, rogetis mortem ipsi aegro." According to the manuscript notes on the
1560–1561 lecture, Falloppia rather cynically added the alternative possibility
of handing the patient over to an "empiric" who would make him or her die
quickly (Wellcome Library, London, Western manuscripts, Ms. 269, fol. 62r).
106 Seidel, Liber morborum incurabilium (1593).
107 Ibid., p. 19.
108 Falloppia, De decoratione (1566).
109 In her *habilitation* on the early history of medical cosmetics, Maria Carla
Gadebusch Bondio (Gadebusch Bondio, Medizinische Ästhetik (2005)) summarized the first chapters of Falloppia's lecture, especially his discussion of
obesity. She wrongly gives the year 1600 (when it was reprinted in the collected
works) as the date of the first publication and, as a result, attributes the lecture
notes to Andrea Marcolino rather than to Agathus who published the work in
1566, before Mercuriale's treatise came out.
110 In his *Canon medicinae*, Avicenna had dealt with some of these conditions but
had focused almost exclusively on their treatment. This also goes for the works
of medieval surgeons like that of William of Saliceto (Saliceto, De decoratione
(1502)), which is largely limited to therapeutic recommendations, especially for
a range of skin conditions.
111 The work first appeared as an appendix to the second edition of Giulio Mancini's lecture notes ("ex ore [...] excepti") on Mercuriale's *De morbis cutaneis* and
De omnibus corporis humani excrementis (Mercuriale, De decoratione (1585)).
In 1587, it was published as a separate work (Mercuriale, Decorationes (1587)).
In his dedicatory epistle, Padua, 15 July 1585 (in the 1587 edition dated 15 July
1586) to Ascanio Piccolomini, Mancini explained that he had turned Mercuriale's public lectures ("explicationibus publicis") into a book. On Mancini and
his surviving lecture notes, see De Renzi, Career (2011); De Renzi (ibid., p. 237)
claims that Mancini oversaw the printing of Mercuriale's work *De decoratione*
(rather than turning his notes on Mercuriale's lecture into a book) but does not
explain what led her to attribute the text to Mercuriale himself rather than to
Mancini. In 1601, a further edition came out with Giunta (Mercuriale, De decoratione (1601)); in 1600, Giovanni Tommaso Minadoi devoted a long treatise
to similar issues (Minadoi, De humani corporis turpitudinibus (1601)).
112 Galen, Thrasybulus (2018).
113 Falloppia, De decoratione (1566), foll. 35r–v, "mulier destruit venustatem: senes
gravitatem amittunt. iuvenes virilitatem perdunt."
114 Mercuriale, De decoratione (1585), foll. 4v–9r.
115 For a detailed overview see Stolberg, "Abhorreas pinguedinem" (2012).
116 Falloppia, De decoratione (1566), foll. 41v–45r.
117 Ibid., foll. 45r–46v.
118 On the long-standing medical interest in keeping the breasts of young women
small and firm see Phillips, Breasts of virgins (2018).

119 This topic was already dealt with by William of Saliceto, De decoratione (1502), foll. 103v–104r.
120 Falloppia, De decoratione, fol. 51v.
121 Ibid., foll. 48r–49r.
122 Tagliacozzi, De curtorum (1597).
123 Falloppia, De decoratione, foll. 49r–v.
124 Falloppia, De vulneribus (1606), p. 368.
125 Falloppia, In Hippocratis librum (1566).
126 Falloppia, Opuscula tria (1569).
127 Falloppia, De parte (1571). According to the frontispiece, this was a second augmented edition but no first edition with the same title is known; presumably, the publishers were referring to the *Opuscula tria* of 1569.
128 Falloppia, De vulneribus (1606).
129 Wightman, Quid sit methodus? (1964); Maclean, Logic (2002).
130 Falloppia, De vulneribus (1606), pp. 241–242.
131 Ibid., p. 245.
132 Ibid., p. 247; the editor added in the margin: "methodus autoris, ad inveniendas differentias."
133 Ibid., p. 252.
134 Ibid., p. 255.
135 Ibid., p. 276.
136 Ibid., p. 276.
137 Ibid., pp. 264–270.
138 Ibid., p. 264.
139 Ibid., p. 317.
140 Ibid., p. 319.
141 Ibid., p. 329.
142 Ibid., p. 351.
143 Ibid., p. 363.
144 Ibid., p. 364.
145 Ibid., p. 258.
146 See the chapter on cosmetic medicine.
147 Falloppia, De vulneribus (1606), p. 372.
148 Ibid., p. 377.
149 Ibid., p. 390.
150 Ibid., p. 393.
151 Ibid., pp. 393–394.
152 Ibid.
153 Ibid., p. 410.
154 Ibid., pp. 408–409.
155 Ibid., p. 407.
156 Falloppia, De luxatis et fractis ossibus (1606), fol. 159v.
157 Falloppia, De luxatis et fractis ossibus (1606), fol. 171r–v.
158 One of the relatively few exceptions is his advice, to loosen the limbs before repositioning a dislocated limb by massaging it with oiled or greased hands rather than with warm water (Falloppia, De luxatis et fractis ossibus (1606), fol. 164r).
159 Falloppia, De luxatis et fractis ossibus (1606), fol. 190v.
160 Heinemann, Henker (1900); Nowosadtko, Wer Leben nimmt (1993); Herzog, Scharfrichterliche Medizin (1994).
161 Falloppia, De luxatis et fractis ossibus (1606), fol. 160r.
162 Ibid., fol. 164v.

4   Materia medica

Falloppia's interest and expertise in surgery, in the treatment of injuries and skin conditions, was exceptional, even for Italian standards. Learned physicians at the time devoted themselves above all – and in the German lands almost exclusively – to the treatment of internal diseases. Pathological processes inside the body might sometimes call for minor surgical interventions such as phlebotomy but the mainstay of treatment was the oral application of all kinds of medicines. An extensive knowledge of pharmacology and a thorough acquaintance with the literally hundreds of plants and other medicinal substances that could be used for the treatment of different diseases was therefore indispensable for any practicing physician.

Even more than anatomy and surgery, studying and teaching medical botany called for both a thorough acquaintance with the ancient writings and the empirical study of nature. The major authority on medicinal plants was far into the sixteenth-century Dioscorides with his *De materia medica* (Περὶ ὕλης ἰατρικῆς).[1] Any serious physician would have to be thoroughly acquainted with this work and Falloppia emphatically recommended that his students read Dioscorides not once only but several times.[2]

In the sixteenth century, physicians began to question some of Dioscorides' descriptions. Moreover, the number of medicinal plants an experienced botanist would ideally be able to identify and distinguish grew dramatically, thanks to the search for new, unknown plants in Europe and the many new, exotic plans that were brought there from far away. By the middle of the sixteenth century, contemporary works on medicinal plants had largely superseded the work of Dioscorides but tellingly the most successful and influential among these new works, that by Pietro Andrea Mattioli, was fashioned as a commentary on Dioscorides.[3]

Medicinal plants

As we have seen, Falloppia began his academic career in Ferrara as a lecturer on the simples. We do not know what qualified him for this position. Quite possibly he had already read the *Materia medica* of Dioscorides as a young man with Gadaldini in Modena. Certainly, he found excellent conditions for

DOI: 10.4324/9781003242000-5

the acquisition of extensive botanical expertise once he studied in Ferrara. His teacher Antonio Musa Brasavola was a leading authority in the field and it was almost certainly on his advice or request that Alfonso d'Este had established a botanical garden in the palace garden.[4] Later, in Padua – and quite possibly also in Pisa – Falloppia gave a range of lectures on the simples and on *materia medica* in general. By then, he was a renowned expert in this field. Andreas Patricius even praised him as a second Theophrastos, the other major ancient botanical authority, next to Dioscorides.[5] Bartolomeo Maranta (1504–1571) asked him for a letter of endorsement for his work on the identification of medicinal plants.[6] Entries, in the student notes on Falloppia's lectures, on plants he had found in various places indicate that he undertook extensive field studies and botanical excursions.[7]

If Falloppia had not died at an early age, he might even have published a major work of his own on medicinal plants. According to his student Marcolino, Falloppia planned to write a commentary on *De historia plantarum* by Theophrastos, perhaps inspired also by the success of Mattioli's commentary on Dioscorides. Theophrastos' work was accessible in Latin translation but Falloppia may well have thought of a commentary on the Greek edition.[8]

Various lectures by Falloppia on different aspects of *materia medica* have come down to us in student notes. In 1565, Andrea Marcolino published his notes on Falloppia's 1557/58 lectures on purgatives under the title *De simplicibus medicamentis purgantibus tractatus*.[9] The edition also offered the printed edition of a long letter – or rather an epistolary treatise – on asparagus, which Falloppia had written to Girolamo Mercuriale in November 1558.[10] In 1566, Petrus Angelus Agathus followed with his notes on Falloppia's commentary on the first book of Dioscorides' *Materia medica*. It was a fragment that dealt with a handful of substances such as *balsamum*, *ambra*, and *zibettum* only;[11] notes on the complete lecture were printed later only, in the *Opera omnia*, under the title *De materia medicinali*.[12] In 1570, the brothers Meietus in Venice came out with *De compositione medicamentorum*.[13]

De materia medicinali dealt with the various medicinal plants and substances, which Dioscorides discussed, one after the other, in the first book of his *De medica materia*. As Falloppia explained, this first book focused on rare and precious aromatic medicines and ointments, many of which were not to be found in Europe. Some of them, like cardamom and cinnamon, are known above all as spices today. Following Dioscorides and widely drawing on other sources as well, Falloppia described each of these medicinal plants or substances. He explained where they could be found, at what time or stage of development they were best collected, what they looked like, and with which other plants (or adulterated, fake medicines) they might be confused. Moreover, he discussed their *facultates*, their powers, and effects on the body, although usually without specifying for which particular kinds of diseases they were suitable. He concluded his lecture with a brief account on civet (*De zibetto*), which had such a pleasant smell that it was topped only be the smell of roasted meat, adding an advice, which, like the *linteolum* against the French disease, clearly was meant to appeal to his male students: when civet was applied to the male foreskin, it would powerfully incite women to intercourse and arouse the greatest pleasure ("delectationem") in them.[14]

De simplicibus medicamentis purgantibus presented a wide range of purgatives that were available to physicians and described their preparation and application. Since sixteenth-century physicians attributed most diseases to some raw, impure, foul, or else burnt, acrid matter in the body, the purgatives played an outstanding role in their practice. In order to fight the disease at its roots, the morbid matter had to be evacuated and "purgantia", which quite literally "cleansed" the body, just as the "purgatory" cleansed the souls, were therefore in most cases the mainstay of medical treatment.[15]

De compositione medicamentorum began with a systematic outline of pharmacology. To compose their own recipes, Falloppia explained to his students, they needed to know, how exactly medicines worked in the body and how their powers could be assessed and put to use. To start with, different types or levels of effect or *alterationes* had to be distinguished. Warming, cooling, moistening, and drying were the first order. The second order of *alterationes* referred to the more specific effects on the human body, such as hardening or loosening, attracting and repelling, the opening or widening and narrowing of pathways, concoction, and fighting pain. Among the *alterationes* of the third order was the regeneration of flesh when wounds healed and the generation of milk and hair. The fourth order was controversial, namely, the effects that were due to some occult "propria substantia".[16] This notion was used, among others, to explain the ability of medicines to attract or repel specifically certain fluids or humors, to promote menstruation, for example, or the evacuation of sweat. It also helped understand why medicines acted more strongly on some parts of the body than on others.

The ability of medicines to work first-order *alterationes* – and especially to warm the body – came in different degrees of intensity. In the case of heat, the effects ranged from a mild sensation of warmth to excruciating pain and burning. The intensity of second-, third-, and fourth-order *alterationes* was far less obvious. There was no clear "latitudo", no identifiable "excessus", regarding looseness ("raritas") and "compactness" ("densitas"), for example. These higher-order *alterationes* resulted from the *temperamentum* of the medicine, Falloppia explained, which was the form that resulted from the mixture of the four elementary qualities. In other words, they were due to its "total substance". The primary qualities were present in the *forma mixti* of these substances not *in actu* but only *in potentia*, which was the reason why the higher-order *alterationes* could only be established a posteriori, by observing the changes these medicines worked in the body.

In one of the most original parts of his lecture, Falloppia then discussed how medicines acted inside the body. To achieve their effects, he explained, minuscule particles of the medicine needed to come into direct contact with and act on the minuscule particles of the body's own substance. This was possible because the body was "perspirabile". It was pervaded, that is, by channels and ducts through which the "minimae particulae medicamenti" reached the "minimas corporis nostri particulas". His students could observe and experience this in the treatment of the French disease with mercury ointment. The ointment was applied to the skin between the thighs, on the spine, and on the legs only and did not seem to penetrate the skin. Yet, when the patient was given a golden ring into the mouth, they would see that it became covered with mercury.[17]

After this general discussion of pharmacology, Falloppia turned to the more specific topic of composite medicines. These medicines differed from the simples in that they did not only consist of one natural substance. When a plant was burnt to ashes, he explained, or when it was distilled, it was still a "simple". By contrast, when other substances were added, it was a "compositum".

The effects of composite medicines were more than just the sum of those of the individual ingredients. Substances like licorice could be added in order to help the other medicines penetrate to the desired location in the body. Some substances helped correct the bad taste, of cassia or rhubarb, for example, or intensified or weakened the effect of a medicine. Honey, sugar, or vinegar protected the medicines against putrefaction. With medicines that had powerful effects in tiny qualities already, making them more voluminous by means of other substances facilitated the oral application, etc.

The bulk of the lecture was then devoted to the rules which guided the composition of medicines, to the interaction of the primary and the higher-order qualities in the mixture and to the different kinds of composite medicines, such as *confectiones, electuaria, trochisci,* and *syrupi.* He concluded with a presentation of some of the major, popular "standard" mixtures that were in use. Some of them, such as *diambra, diacalamentum, diaciminum,* and *diagalanga,* could be found already in Mesue and other Arabic writers. Others, like the *confectio cordialis,* were of recent invention. One of the most widely used composite medicines was *manus Christi perlata,* a sugary medicine made of various plants to which finely ground pearls were added. According to Falloppia, the origin of the name was unclear; maybe it referred to the powerful beneficial effects, which were likened to those of the hand of Christ. The country folks ("rustici"), he claimed, held such a strong belief in the powers of *manus Christi* that thought they would die without it.[18]

Under the title *De medicamentis simplicibus,* later editions of the *Opera omnia*[19] added the notes of an unknown student on Falloppia's introduction to the general theory of faculties of different types of medicines, which preceded his lectures on Dioscorides.[20] Compared to his discussion of similar questions in *De compositione medicamentorum,* these lectures give the impression of an overly theoretical and somewhat disorganized account, which was of limited use to future practitioners. After an extensive discussion of the precise definition of "simples" and "faculties", Falloppia divided medicinal substances into four *genera.* The first group were medicines like *cicuta* and *mandragora* that worked by their sheer quantity rather than by their total substance. By contrast, medicines that belonged to the other groups had to undergo a change ("alteratio") in the body to develop their powers. Those of the second *genus* were harmful and poisonous, those of the third effected changes in the body without harming it, and those of the fourth group not only did not harm the body but also nourished it. It seems that his students got understandably impatient with this rather convoluted analysis. Falloppia felt compelled to defend himself against those who complained and wondered why he spent so much time on "universals".[21]

In his various lectures on *materia medica*, Falloppia underlined the importance of empirical knowledge in this field. The crucial place of observation even led him to attribute a certain value to the knowledge and experience of ordinary folks. He placed them in a hierarchy of botanical expertise. The *herbolarij*, who brought medicinal herbs from the mountains into the towns, knew where to find and how to identify and distinguish the different plants with their respective names but nothing else. The pharmacists ("seplasiarij") were more knowledgeable. They had also learned how to preserve and make use of the different substances according to the physicians' instructions. Some ordinary women ("mulierculae", "vetulae") also had some insights into empirical matters ("in empiricis") but they used medicines based on experience only without considering the nature and the *temperamentum* of the individual plant. The rational physician alone possessed the full, complete knowledge that was necessary for a medical practitioner. He knew the names of the different plants and what they looked like as well as their nature and their powers ("facultates", "vires"), which allowed him to choose medicines that preserved health and drove out diseases.[22]

Falloppia accordingly stressed, on the one hand, the need for the learned physician to acquire extensive book knowledge in this field and advised his students to read the relevant authors. He himself quoted a wide range of works, those of Galen, Dioscorides, Theophrastos, and Pliny, in particular, but also those of Arabic writers like Rhazes and Averroes and of contemporary ones such as Pietro Andrea Mattioli, Amatus Lusitanus, and, for different types of preparations, Valerius Cordus.[23] At the same time, he admonished his students to study the medicinal plants in vivo, in the botanical garden and roaming the countryside.[24]

The University of Padua had a botanical garden since 1545, which in Falloppia's days was directed by Aluigi Anguillara.[25] It is not clear to what extent Falloppia personally instructed his students on the various plants, their different stages of development, the shape of their leaves, flowers, and roots, and their medicinal preparation in the garden. A letter from the *Riformatori* in Padua to the *Capitano* in Padua suggests that the professors of practical medicine also taught there: the *Riformatori* complained about some students who were not content with looking at the plants in the botanical garden but dug out many herbs and even insulted the guards. They asked the *Capitano* to tell Vittore Trincavella and Antonio Fracanzano – both were professors of *medicina practica* – and Gabrielle Falloppia that they should admonish the students accordingly.[26] Georg Handsch compiled long lists of the medicinal plants he saw in the botanical garden[27] but did not explicitly mention Falloppia in this context.[28] He only described visits with Fracanzano to the local apothecaries to learn about the various medicinal plants and their application.[29]

Metals, stones, and earths

Medicinal plants were at the center of teaching *materia medica* but there were also other areas on which Falloppia taught extensively. In 1556 and 1557, he lectured on mineral waters and thermal springs and on metals and minerals, drawing extensively on the works of ancient authorities. The two lectures are documented in the detailed notes, which Falloppia's student Andrea Marcolino published in 1564 under the title *De medicatis aquis atque de fossilibus*.[30] *De medicatis aquis*, later also known as *De thermalibus aquis*, dealt with the origin, the nature, and the curative effects of thermal waters in general and then described a whole series of springs, mainly in the area around Padua and in Tuscany.[31] *De fossilibus* or, as the relevant section is more precisely titled, *De metallis seu fossilibus tractatus* dealt only marginally with fossils in the modern sense. The principal topic here were metals and minerals and their medical uses. Marcolino maintained the style of a lecture, with Falloppia addressing the students with "you" and rendering his personal observations and opinions in the first person. As Marcolino explained by way of introduction, he also kept Falloppia's arrangement. He only omitted the repetitions of the material of the preceding lecture with which Falloppia began each lecture, inserted headings, added bibliographical references in the margins, and prepared a detailed index.[32]

The lectures began with a detailed natural-philosophical discussion of the nature and genesis of the waters, metals, and minerals in the earth's interior, weighing opinions and counter-opinions in a scholastic manner. Falloppia made it clear, however, that he dealt with these topics as part of his teaching on *materia medica* or, to use the more general modern term, "pharmacology". He explicitly referred to a preceding lecture, in which he had discussed the last part of the fifth book of Dioscorides' work that was devoted to stones and metals.[33] The division into the two major thematic blocks also went back to Dioscorides. Dioscorides, Falloppia reminded his audience, had distinguished two types of matter which were generated in the bowels of the earth, namely those that came out on their own, for example, in the form of steam, smoke, water, and those that had to be mined.[34]

We will first look at his lectures on metals and minerals, which under the title *De metallis seu fossilibus tractatus* occupy the second part of the volume of 1564.[35] Here Falloppia first discussed in a scholastic manner the different opinions of the authorities, quoting extensively the works of the ancient philosophers. Aristotle, Plato. Dioscorides and Galen and, among the recent authors, Georg Agricola were his principal authorities. Falloppia distinguished three types of metals and minerals, namely the earths ("terrae"), the stones ("lapides"), and the metals ("metalli"). Drawing on Galen and Agricola, he initially added a fourth genus, the "juices" ("succi") and the salts, gems, and the like that originated from their solidification.[36] Later in the course of his lecture, he no longer conceded a separate status to the "juices". Ultimately, he explained, they were only the fluid parts that were eventually absorbed by earths and stones.[37]

For each of the three genera, he discussed their nature and definition, their material cause, their origin, and the efficient cause that produced them. "Terra", he made clear, did not refer to the element of the ancient natural philosophers, which no one had actually ever seen in its pure form but to ordinary soil. He rejected Agricola's opinion that the different kinds of soil were formed from different stones. Soil, he argued, originated from smoke and vapors that arose from the hot interior of the earth, where sulfur or bitumen were burning and produced black smoke, which solidified when it cooled off. The differences between various types of earth were due to the respective matter from which the vapors or smoke originated, the intensity of the heat that acted on it, the types of matter through which vapors passed on their way to the surface, and the respective mix of smoke and steam.[38]

Stones differed from metals in that they did not liquefy under the influence of heat. Some authors disputed this and said that stones could also melt. If experience seemed to confirm their claim to some extent, this was only because some fluid was mixed in. Stones themselves did not liquefy.[39] Falloppia distinguished four types of stones that occurred in nature, apart from man-made stones such as clay bricks. They all originated from a "stony juice" ("succus lapideus"). Their genesis could be observed in places where water emerged from the earth, bringing forth stones that grew bigger and bigger with time.[40] Such a stony juice also explained how stones could originate from plants or wood – clearly Falloppia was referring here to "fossils" in the modern sense. In Venice, he told his students, he had seen a large oak branch transformed into a beautiful snake-like stone. Corals and other plants in the sea formed stones by attracting stony fluid from the rocks in the depths of the sea, which then hardened and petrified.

A third group were stones that formed in living things. Some were pathological and of particular interest to the medical practitioner, like the stones found in the gall bladder and in the various organs of the human body. They resulted from a viscous fluid that solidified under the impact of heat. Other stones formed in animals due to their peculiar nature. Oyster pearls were nothing but stones that originated from a very pure stony juice, which the oysters attracted from the rocks to which they clung. They boiled this juice so exquisitely that extremely pure and beautiful stones were formed – the pearls. The existence of such a stony juice in the seawater was proven by the stony cones that the students could see on the pillars of the bridges in Venice.[41]

The last and biggest group were the stones that were formed in the earth and in the "bowels" ("viscera") of the mountains, including the crystals or precious stones ("gemmae"). Falloppia rejected the "vain opinion" of Cardano that stones nourished themselves, grew, and repaired defects.[42] To do so, they would have to be able to attract stony juice and distribute it evenly, like the human body did with food and medicines – but small stones had no veins or crevices.

Turning to the metals, Falloppia devoted even more time to theoretical issues and natural philosophy, presenting and rejecting the views of a range of authors on the origin and the *materia* of metals.[43] Referring to the *chimistae*, he went to considerable lengths to reject the theory that all metals were made of mercury, sulfur, or both, which gave legitimacy to their claim that gold could be made from other metals. The chymical art could look back at a long tradition and some highly learned men had practiced it and Falloppia listed various arguments for their claim: hard metals could not originate from fine vapors – there had to be something intermediate, like mercury. The same way, in which the female was attracted by the male as something like its perfection, mercury was attracted by metals, to which it adhered, and in particular by gold. Moreover, the *chimistae* claimed that gold and silver could be made from mercury. A pharmacist from Tarvisio, known as "il spetiale dal Saracino", even was reported to have given proof of his ability to make gold from mercury in Venice in front of the entire senate. Yet, the claims about mercury as the matter of metals and the possibility of making gold and silver from it were false. The pharmacist from Tarvisio had deluded his spectators, and he had been punished for it. It was true that metals were rarely pure and it was therefore possible that gold and silver could be extracted from other metals, like mercury. The quantities were tiny, however, a minuscule fraction, and extracting gold was not equivalent to making it.[44]

A highly controversial question, which Falloppia discussed in depth in this context, was the question whether stones had a soul, whether they were animate. In the late 1550s, when Falloppia delivered his lectures, the reception of Paracelsian medicine in Italy was just beginning on a broader scale. Given the influx of students and physicians from the German-speaking world to Germany, we may assume that Falloppia was at least roughly informed about Paracelsian doctrine. When Thomas Erastus, a staunch opponent of the therapeutic use of metals, published his detailed critique of the "new theophrastic medicine" in 1572, he even explicitly referred to Falloppia's lecture, in his discussion of the nature of metals. Falloppia and Julius Caesar Scaliger, he found, had said everything that could be said on the subject.[45] In his lecture, Falloppia did not discuss the Paracelsian ideas as such. His criticism was directed against Neoplatonic currents in recent natural philosophy in general and against Girolamo Cardano (1501–1576) in particular, who ranked stones among the living beings and claimed that metals grew in the mountains.[46]

The central argument for the assumption that metals and minerals were animate was the widely accepted experience that stones and minerals "grew" in the earth, and sometimes even did so in forms similar to the vessels in the animal body or the roots of plants. Moreover, when stones were removed from the surface in one place, other stones were seen to grow in their place and when stones were damaged, the "wound" healed or "scarred" over time. All this was taken to indicate that the stones and minerals were able to attract and assimilate nutrients, which, it was widely agreed, required the action of a vegetative soul. Falloppia disagreed. He did not doubt that stones and minerals grew back and that damaged stones "healed" but he had a different explanation. The growth did not result from active nutrition, which would be the work of a soul, as in plants and animals. Growing and healing were due to the mere aggregation of matter from the outside. That this was the case could sometimes be seen from the various layers of matter that formed.[47]

Marcolino added to his notes that Falloppia did not have enough time in the end to bring his extensive theoretical discussion to a proper conclusion. Provided Marcolino's notes are complete, Falloppia jumped rather abruptly, in fact, to the individual metals and other substances which could be mined and could be used to treat diseases, including cadmium, molybdenum, iron in its various forms, lead, tin, antimony, mercury, and cinnabar, as well as non-metallic matter like ochre, lapis lazuli, and the *lapis* or *bolus armenus*, which he praised as a God-given medicine among others against the plague.[48]

In between and somewhat out of context, Falloppia once more offered his male students also a special piece of knowledge of potential personal interest. Having pointed out the emetic and laxative effects of *chrysocolla*, he added a brief remark on the special powers of *borax officinalis*: he told his students about a Tuscan physician who gave a scruple of powdered borax in a decoction of a warming substance such as *matricaria* or *sabina* (savin) to women who could not give birth. The medicine performed miracles, expelling the fetus within a short period of time, dead or alive. It must have been clear to everyone present that this kind of medicine could be very welcome, in case one of his students made an unmarried woman pregnant.[49]

Thermal springs and mineral waters

The medical use of thermal springs has a long tradition in southern Europe, dating back at least to Roman times. In the Middle Ages, visits to baths and thermal springs for the treatment of illnesses became increasingly popular again, especially in Italy and a trade in mineral waters started to develop. According to Falloppia, the waters from the Bagni di Villa near Lucca, for example, could be bought almost everywhere in Italy, except in Venice, and they were also exported to other countries.[50] Since the fourteenth century, learned physicians such as Giovanni de Dondi (ca 1330–1388)[51] and Giovanni Michele Savonarola (1385–ca 1466)[52] had written extensively about thermal springs. In 1553, the Venetian publisher Tommaso Giunta edited a volume on the topic with various texts by ancient and recent authors, including those of Dondi and Savonarola.[53] In his lectures, Falloppia also mentioned a treatise by his Paduan colleague Francesco Frigimelica on mineral waters, in general, and told the students to ask Frigimelica to put it into print.[54] No such work seems to have been published during Frigimelica's lifetime but it may have circulated in a manuscript. In 1659 and possibly based on the manuscript Falloppia had in mind, a treatise by Frigimelica on the artificial preparation of medicinal waters by adding metals appeared, which also discussed mineral waters and thermal springs in general.[55]

The topic of thermal springs and mineral waters was of particular relevance in Padua. There were various well-known thermal springs in the area. Those in Abano Terme are still famous today. As Falloppia told his students, the waters from Abano had once even been brought by means of an aqueduct to a public bath in Padua and the remains of that aqueduct could still be seen in some places.[56] As a medical practitioner, Falloppia like the other physicians in Padua greatly appreciated the medicinal value of thermal springs and recommended their use for the treatment of all kinds of diseases. Among the about forty documented cases of patients on which Falloppia expressed his opinion, orally or in writing, there are hardly any, in fact, in which he did not also recommend a suitable thermal spring.[57]

The term "thermal" had to be taken with a grain of salt, Falloppia explained. Most of these waters were indeed warm or hot when they emerged at the surface but in some places the mineral water that issued from the earth was actually cold: it had originally been hot but had cooled off during its passage from the earth's interior to the surface.

Falloppia discussed at length and rejected various theories that had been proposed since ancient times to explain the heat of thermal waters.[58] The sun could not be the cause. The sunrays could penetrate loose earth at best but not the rocks from which many thermal springs sprang. Moreover, the water from thermal springs was often far hotter than soil that was warmed by sunlight. The winds or the friction of the flowing cold water against rocks could not generate a sufficiently strong heat either. There had to be a permanent fire inside the earth, which heated the water.

This raised the question of the material cause of this fire, of the substrate or fuel that maintained it. Democritus' explanation that the heat came from lime and ashes might seem plausible, at first glance, Falloppia argued. Both substances were often found in the vicinity of thermal springs and experience showed that intense heat developed when lime was mixed with cold water. This heat was only momentary, however. There had to be another explanation. According to the (Pseudo-)Aristotelian *Problemata*, the fire in the earth's interior was maintained by burning sulfur, while Georg Agricola identified bitumen as its fuel.[59] Falloppia combined both positions: many large and small fires burned beneath the earth, some fueled by sulfur, others by bitumen, and others again by both. Of course, the sulfur and the bitumen were consumed with time but experience showed that the earth constantly generated new sulfur and bitumen: when a sulfur mine was not used for a couple of years, the miners would find it full of sulfur again when they re-turned. The same was true for bitumen, as could be seen in the area around Modena.[60]

Depending on the strength of the fire and the distance the water passed through the rocks before it reached the surface, it could be burning hot, lukewarm, or cold. More importantly, depending on the kind of matter they carried with them during their passage through the rocks, waters from dif-ferent thermal springs had very different effects. The practicing physician therefore had to be familiar with the composition and the qualities and fac-ulties of different waters. On this basis only, he could recommend a specific thermal spring for certain ailments or indeed for the individual patient who asked for his advice.

Unfortunately, Falloppia explained, the ancients, who had written so care-fully about other topics, were not very helpful here. Much of what had been written about different healing waters smelled of "empyria".[61] All kinds of phantastical qualities were attributed to the various waters, and often au-thors just repeated what they had read in the works of others, without hav-ing even seen the springs themselves. Falloppia contrasted his own method for the investigation of healing waters. Frequently, important clues could be obtained from the senses, from the smell, the taste, the visual appearance, and the way the water felt on the skin. The problem was that the senses were easily deceived, however. The only method that produced certain knowledge was similar – though Falloppia did not explicitly make that comparison – to the dissection of corpses, which led to certain knowledge of the human body and its parts. It was laborious, which was why writers had preferred to simply follow, like sheep, those who had written before them and perpetu-ated their errors. The water had to be "dissolved" into its various constituent parts – Falloppia even spoke of *minima particula* in this context.

This analysis, as we would call it today, could be undertaken in different ways. The water could be boiled or simply left standing to let it evaporate gradually. When the water was poured into a wooden vessel with a little crack through which it could leak, salty concretions could be found around the crack the next morning.[62] The best, most certain method by far, however, was distillation or "elambichatio" as Falloppia called it with a term that was closely related to the more familiar term "alembic", which was commonly used for a distilling apparatus. Falloppia recommended and described the use of a simple distilling oven for this purpose. In the printed edition, the device is shown and explained in an illustration (Figure 4.1).

A urine glass or some similar kind of glass vessel ("bocia") with the water was put into an oven chamber that was made of burnt clay and filled with sand. The chamber and the vessel which was placed above it were heated by a fire from underneath. The heat made vapors gradually ascend from the water. They were caught by a hemispheric *capitellum*, turned fluid again, and flowed toward the outside through a long glass tube ("fistula"). On its way, this glass tube passed through a vessel full of cold water that further cooled

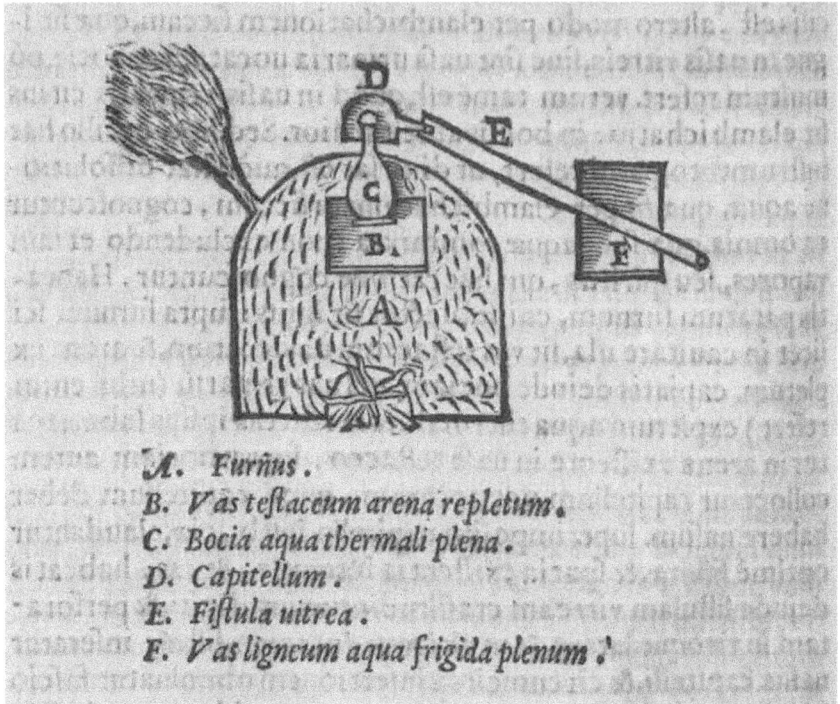

Figure 4.1 Apparatus for the distillation of mineral waters, in: Falloppia, De aquis (1564), fol. 35v.

the fluid before it exited from the open end of the tube where its different fractions could be collected and examined. When all the water had evaporated, the glass vessel was removed from the furnace and the remaining sediment was examined. Certain earths, such as ocher, had a specific smell. By rubbing the sediment between the fingertips, certain admixtures like sulfur, orpiment, and the like could often be felt. Then the sediment had to be dried and examined in the sunlight. Sometimes shiny corpuscles could be seen, suggesting the presence of salt or niter. Sulfur could be distinguished by its peculiar color. The presence of other components such as gold, silver, tin, and iron could not be ascertained in this manner. The taste when they were put into the mouth also varied. To identify and distinguish them one had to sprinkle some of the dried sediment on a clean glowing-hot iron. Alum sometimes but not always melted and became white as milk. Salt and niter sparkled but only salt also produced a crackling sound. Lime, marble, and gypsum did not burn but turned white; in the case of gypsum that happened faster than with the other two and the white was more intense. Sulfur produced a characteristic smell under the influence of the heat. Falloppia had not been able to find lead in the water this way but he believed that it would melt on the hot iron. The presence of lead white gave the sediment a reddish color.[63]

Sometimes Falloppia also added substances, which, by the changes they effected, indicated the presence of certain metals or other substances in the water. Those who claimed the water in the Bagni di Calderiano near Verona contained iron rust ("ferrugo") were mistaken, he found, for example. He had examined them "by all those means, by which red ocher ('rubrica') was distinguished from rust" adding drops of vinegar and adstringent substances and using all the other methods he had explained before and there clearly was no rust.[64]

Falloppia's method for the chemical analysis of mineral waters exerted a considerable influence on the many early modern physicians and chemists after him who devoted themselves to the study of mineral waters, thanks also to Conrad Gessner's *Euonymus*, which offered a detailed description of the way in which he proceeded.[65]

After this extensive natural-philosophical and methodological exposition, Falloppia provided detailed accounts of a number of thermal springs he had personally inspected and examined. He described those in Abano and in other nearby places but also some well-known springs further away, near Verona, Reggio, Lucca, Pisa, and Volterra. Some of them he knew from his days in Pisa or indeed from his childhood years in Modena. He described the various fountains and wells that could be found in these places, the tubs and mud baths, and the buildings and their natural surroundings. He told his students for what kinds of ailments the locals and other visitors used these waters and how they were applied. Sometimes he also commented on the social life in these places. In a thermal spring near Pisa, for example, he found the sick spending hours in the water and singing various songs to keep up their good spirits, some modest and others less respectable, and telling countless stories and fables.[66] At the center of his account, however, were the contents of the waters in the various springs, which he had identified by distillation. Sometimes knowing them was already sufficient to suggest their use in certain diseases, for example, when the substances he identified in the water were known to have a relaxing or else fortifying, adstringent effect. He combined the result of his "chemical" analysis with his personal observations on the curative effects of the various waters on patients. By insisting that he himself had "experienced" ("expertus sum") the beneficial effects of the waters from this or that spring on patients with certain complaints, he implicitly ascribed a higher degree of certainty to his conclusions: they were more reliable than what earlier authors or the locals reported often from hearsay only. Like other physicians at the time,[67] he sometimes also resorted to an even more reliable source of knowledge: his own body. Describing the fourteen *stillicidia* in the Bagni di Corsena, where patients let very hot water drip on their heads, he expressed his gratitude to God who created them. He had experienced their curative effects twice, he pointed out, at the age of fourteen and then at the age of twenty eight, when he was almost deaf and regained three-fourths of his lost hearing thanks to them. Like him, others recovered from deafness in Corsena and some also regained their lost eye vision.[68]

Empirical observation: experimentum, periculum, and total substance

Historians have highlighted a growing appreciation for observational knowledge in the sixteenth century, for "autopsia" in the literal sense of "seeing oneself" especially with regard to the rise of anatomy. This was a field in which Falloppia excelled and where he contributed a range of new findings. Botany, the identification of the plants described by the ancients and the discovery and study of domestic and exotic plants unknown to the ancients, is another well-studied area in which experience and the personal observation played a crucial role. Historians have so far seriously underestimated, by contrast, the rise of empirical approaches – long before the so-called scientific revolution of the seventeenth century – in an area that was at the very core of medicine and the medical profession: medical practice. It was above all the quest for more efficient remedies and for better therapeutic outcomes that made many physicians turn toward a growing reliance on empirical observation.[69]

Falloppia's research on the contents and curative properties of thermal springs and their mineral waters is a prime example of this trend and of the observational terminology that came with it. As Marcolino's notes indicate, Falloppia frequently used the expression "periculum facere", that is, literally "making a little experience" or "trial", to refer to his examination and observation of the particular properties of the various waters. "Feci omne periculum", he said, for example, about the waters of the Bagni di Villa near Lucca, which some claimed contained iron but in which he could not find any iron.[70] "Feci pericula omnia", he rejected the claim that the waters of San Giovanni di Corsena contained alum. He could not find a trace of alum, just salt and large amounts of *nitrum* and *calx* and, despite the high temperature, only little sulfurous vapor, which suggested that the water was heated on its way to the surface by rocks under which sulfur was burning.[71] Indeed, he only wanted to give his opinion on those waters, which he had experienced himself ("ipse expertus sum") and remain silent on those of which he had not made a *periculum* by using them on patients and examining the substances they contained.[72]

As Evan Ragland has shown, the expression "facere periculum" became quite common among physicians during the sixteenth century. In the case of the famous and influential Falloppia, Ragland finds, "facere pericula" served in particular to reject the claims of others.[73] By contrast, when Falloppia used the expression "facere experimentum", he had, according to Ragland, no specific claim or thesis in mind.[74] I do not quite agree with

Ragland's otherwise enlightening analysis on this second point. The term "experimentum" – this Ragland does not mention – was widely used among sixteenth-century physicians more specifically to refer to a remedy that had been found to be efficacious. In this sense, medical practitioners collected countless recipes for "proven" remedies as "experimenta" in their notebooks.[75] According to the student notes, Falloppia praised his own "experimentum" ("proprium experimentum") accordingly, which he had given thirty times and mostly with success during the plague epidemic in Padua in 1555 and he considered it unwise to change the composition of a "maximum experimentum" against the plague.[76] Falloppia sometimes did use the expression "experimentum facere" in a general sense, to refer to the examination of substances like the mineral waters,[77] but in his usage, "experimentum facere" referred primarily to making an observation that successfully demonstrated the therapeutic virtues of a certain medicine on patients.[78] In this sense, Falloppia claimed, for example, that he saw and experienced ("feci experimentum") the outstanding therapeutic effects of the bezoar stone as an antidote against the plague poison on several Portuguese patients in Ferrara.[79] In other words, "periculum facere" was equivalent to "making a trial", without knowing its outcome (and sometimes to prove one's point against others) while "experimentum facere" referred primarily to the successful empirical demonstration of the previously assumed therapeutic effects of a medicine.

Some historians have even praised Falloppia for performing one of the first true pharmacological experiments in the history of Western medicine.[80] The claim is based on the aforementioned story, which Falloppia repeatedly told his students, about the men who were sentenced to death and sent to Pisa to be anatomized by Falloppia and to one of whom Falloppia first gave some opium during a fever attack, which he survived, and then a second dose between two fever paroxysms, which proved fatal. As we have seen, Falloppia did not administer the opium to study its effects, however. It was a mishap. Falloppia wanted to kill the man in order to dissect him afterward and found, to his surprise, that he survived the first dose that was administered during a fever paroxysm. It was only in retrospect that Falloppia concluded that the febrile heat of the paroxysm had counteracted the effects of the deadly cold opium.[81]

An entry in one of the personal notebooks of Falloppia's student Georg Handsch comes somewhat closer to describing an experimental setup, even though the medicine was again not administered for the purpose of observing its effects. Handsch not only mentioned the story of the man with a quartan fever but also documented Falloppia's account of the effects of two drachms each of opium on altogether nine people that had been sentenced to death and sent to Pisa to be dissected. What he saw, Falloppia explained to his students, proved Dioscorides' description of the effects of opium wrong. They all sweated but "none of those symptoms followed which Dioscorides describes".[82] Handsch's account even suggests a modest – and ultimately futile – effort at quantification: "It was said to be two drachms"; he added, "but they did not drink it all, because the greater part was in the dregs."[83]

The move toward an empirical assessment of drug effects went hand in hand with and was promoted by a marked shift in the theoretical foundations of learned pharmacology. Plants and other medicinal substances continued to be classified according to their primary qualities – warm, cold, moist, and dry – and their secondary qualities – such as relaxing, softening, and adstringent – that were believed to derive from them. The deadly effects of opium, for example, were traditionally attributed to its intense cold, which overcame the vital heat. This was also Falloppia's implicit explanation for the survival of the man with a paroxysm of quartan fever. However, even the most orthodox Galenist physicians among Falloppia's contemporaries accepted that certain medicines and poisons worked by some hidden, supralementary quality. Especially the powers of substances, minuscule quantities of which had strong or indeed deadly effects on the human body, could only be explained in this way, the poison of plague and the French disease, for example, as that of snakes and scorpions. In this sense, Falloppia, like other physicians of his time, attributed the effects of "specific" medicines, like the so-called lemnic earth, and of poisons, including poisonous disease matter, to an occult property, to their "propria natura", or, with a term that already played an important role in Galen's work, to their "tota substantia".[84]

The central place physicians like Falloppia attributed to "occult qualities" and the "tota substantia" had far-reaching epistemological consequences. The effects of medicines and poisons that resulted from their primary and secondary qualities could be derived rationally from their known *temperies*, physicians believed. By contrast, the effects of medicines and poisons that were due to occult properties or their "total substance" could only be determined empirically, by the repeated observation of their beneficial effects against specific symptoms or ailments. The sometimes miraculous effects of bezoar, Armenian earth, unicorn powder, and other "proven" specifics against the plague and other fatal diseases could not be derived from their peculiar mix of primary qualities. They could only be known from experience.

Their appreciation for empirical observation also made some physicians take the experiences of their patients and of the common folks in general more seriously, in spite of the common denunciation, in medical writings, of the "ignorance" and "superstition" of the *vulgus*. Falloppia was one of them. Once he told his students a story from his own childhood: as a boy, he saw a young farmer in the fields who was bitten in the toe by a viper. The young man's father tied the toe tightly with a dog leash so that it swelled and turned dark, and cut the skin so that much blood flowed out. Then he told the son to put his foot on the spot where the snake had bitten him. He marked the spot and loosened the earth there while reciting the Lord's prayer. He put some of the loosened earth in wine and made his son drink it. The son vomited violently and had massive diarrhea – and recovered. When the doctors in Modena heard about it, they thought that the vomiting and diarrhea had freed the son from the poison. Falloppia, however, had another explanation: the earth, he explained to his students, had worked through an "occulta proprietas", its "propria substantia".[85] He left it open, whether this "occulta proprietas" was due to or at least enhanced by the little ritual they performed.

Falloppia also told his students the story of a noblewoman in Lucca who called him for advice. She had just given birth to a child with a "hare face", that is with a nose like a hare, with hair on it. She knew the cause, she believed: in her pregnancy, she had seen a boy who carried a hare and was taken by a great desire to eat hare and feeling an itch, she also scratched her nose. She bought a hare and ate it for dinner. As Falloppia explained this was a common belief among the "women" that when they strongly imagined something and rubbed some part of their body the child would receive a "sign" of it, through the force of imagination. He had asked various women and they all had confirmed that this was true. In his lecture, he did not explicitly reject the idea.[86]

Falloppia also reported having seen several times how women who claimed that they could drive the bad spirits out of the bodies of bewitched women performed a cure. They gave a strong dose of white hellebore to these "enchanted" ("incantatis") or "inspired" ("inspiratis") women. Falloppia warned the students of the dangers of this treatment. These women seemed like dead after the massive vomiting provoked by the hellebore. He did not express doubts, however, that the cure had been successful.[87]

The reliance on personal experience and on empirical observations, one's own and those of others, including even those of less learned contemporaries, held certain risks for the professional self-fashioning of the learned physicians. It brought them dangerously close to the numerous lay healers, the "empirici" or "empyrici" as they continued to be called because they "only" relied on experience. However, the potential rewards – better therapeutic outcomes – outweighed the reservations. In his lecture on the plague, Falloppia even explicitly put experience above reasoning: the bad thing about reasoning ("ratio") was that it could see things differently and that ultimately the one who had more authorities on his side and was more eloquent was most successful at convincing others. When human life was in danger, the proven efficacy of a medicine ("experimentum") was the primary guide and one should trust experience more than reason.[88]

Notes

1 Dioscorides, De medica materia (1547).
2 Falloppia, De compositione (1570), fol. 21r.
3 Mattioli, Commentarii (1554).
4 Nutton, Rise (1997), pp. 11–15.

5 Patricius, Dedicatory epistle (1558), fol. 4r.
6 Letter from Falloppia to Bartolomeo Maranta, Padua, 3 August 1558, in: Maranta, Methodi (1559).
7 In September 1554, Falloppia mentioned plans for a botanical field trip together with Wieland (Biblioteca Universitaria di Bologna, Mss Aldrovandi, 38², I, fol. 44, letter from Falloppia to Ulisse Aldrovandi, Padua, 20 September 1554, ed. in Di Pietro, Epistolario (1970), pp. 25–26).
8 Dedicatory epistle to Cardinal Aloysius d'Este in Falloppia, De simplicibus medicamentis purgantibus (1565); cf. Theophrastos, De historia plantarum (1529); idem, Περὶ φυτῶν ἱστορία (1552).
9 Falloppia, De simplicibus medicamentis purgantibus (1565); Falloppia concluded the lecture on 18 July 1558 (ibid., p. 253).
10 Letter to Girolamo Mercuriale, 1 November 1558, published as *De asparagis* in Falloppia, De simplicibus medicamentibus purgantibus (1565), pp. 254–263.
11 Falloppia, Opuscula (1565), foll. 23v–33v.
12 Falloppia, De materia medicinali in lib. I Dioscoridis, in idem, Opera (1600), vol. 2, pp. 25–59.
13 Falloppia, De compositione medicamentorum (1570).
14 Falloppia, De materia medicinali (1606), p. 248.
15 Stolberg, Learned physicians (2022), esp. pp. 183–189.
16 Falloppia, De compositione (1570), fol. 6r.
17 Ibid., foll. 17r–18r.
18 Ibid., fol. 59r–v.
19 Falloppia, De medicamentis simplicibus (1600).
20 At the end of the lecture, Falloppia announced that he would proceed to discuss the first book of Dioscorides' work (Falloppia, De medicamentis simplicibus (1606), p. 210.
21 Falloppia, De medicamentis simplicibus (1606), p. 192.
22 Ibid., p. 186.
23 Cordus, Pharmacorum (1546).
24 Falloppia, De compositione (1570), fol. 21r: "oportet, ut per campos, hortos, et alia loca peragretis."
25 Minelli, L'orto botanico (1995); on Anguillara see De Toni, Luigi Anguillara (1921).
26 ASV, Venice, Riformatori 63, letter from the *Riformatori* to the *Capitano* in Padua, 12 May 1555.
27 ÖNBW, Cod. 11210, foll. 115r–120v: "Herbae quas didici in horto Paduano."
28 Ibid., fol. 119v, Handsch mentions a botanical excursion he undertook with several others, including a "Gabriel"; this was how he sometimes referred to Falloppia but he also had a fellow student by the name of Gabriel Hummelberger.
29 Ibid., fol. 140r.
30 Falloppia, De aquis (1564). Two identical editions of this work appeared in 1564 but with slightly different frontispieces. One of them not only gives the name of the publisher, Aranzi, but also that of the printer, "Ex officina Stellae Iordanis Ziletti". In 1569, the same publisher, Aranzi, published a further, by all appearances identical edition. On the fortunes of this work, see also Ferrari, L'opera idro-termale (1985).
31 Falloppia, De aquis (1564), foll. 3r–85r; for the dating see ibid., fol. 85r ("Patavii die nona Iulii, MDLVI").
32 Andrea Marcolino, letter to the reader in Falloppia, De aquis (1564).
33 Falloppia, De aquis (1564), fol. 85v.
34 Ibid.
35 Ibid., foll. 85v–176r.
36 Ibid., fol. 86r–v.

37 Ibid., fol. 93r.
38 Ibid., foll. 87r–91v.
39 Ibid., fol. 94v.
40 Ibid., fol. 97r–v.
41 Ibid., 97v.
42 Ibid., 101v.
43 Ibid., foll. 111v–124v.
44 Ibid., foll. 119r–124v.
45 Erastus, De medicina nova (1572), p. 257.
46 Cardano, De subtilitate (1582), pp. 260–262 and p. 338.
47 Falloppia, De aquis (1564), fol. 103r–v.
48 Ibid., fol. 159r.
49 Ibid., fol. 158r–v.
50 Ibid., fol. 82r–v.
51 Dondi, De fontibus (1553).
52 Savonarola, De balneis (1485).
53 Giunta, De balneis (1553).
54 Falloppia, De aquis (1564), fol. 3v.
55 Frigimelica, De balneis (1679), 1st edn. 1659; the printed text was based on a
    copy in the library of the Danish physician Johannes Rhodius (1587–1659), who
    obtained his doctoral degree in Padua.
56 Falloppia, De aquis (1564), fol. 4r.
57 See the chapter on medical practice.
58 See also the summary of Falloppia's discussion of the various opinions in Hsu,
    Gabriele Falloppia's "De medicatis aquis" (1993), pp. 79–86.
59 For a detailed discussion of the these controversies see Vermij, Changing theo-
    ries (1998).
60 Ibid., foll. 15v–16v.
61 Ibid., fol. 28v, "earum scientia et cognitio sapiat empyriam."
62 Ibid., fol. 81r.
63 Ibid., foll. 35v–37v.
64 Ibid., fol. 80r.
65 Gessner, Euonymus (1569), foll. 24r–27v, "Modus destillandi aquam sim-
    plicem, & aquas fontium thermalium […] ex Gabrielis Fallopij lib. de aquis med-
    icatis"; see also Debus, Chemical philosophy (1977), pp. 17–18.
66 Falloppia, De aquis (1564), foll. 83v–84r.
67 See the chapter on "Self-observation. The physician's body as a source of knowl-
    edge" in Stolberg, Learned physicians (2022), pp. 390–392.
68 Falloppia, De aquis (1564), foll 81v–82r, "ego sum bis in me ipso expertus".
69 For a more detailed study of this trend see Stolberg, Empiricism (2013) and idem,
    Learned physicians (2022), pp. 357–405.
70 Falloppia, De aquis (1564), fol. 29r.
71 Ibid., fol. 81v.
72 Ibid., fol. 84v.
73 Ragland, "Making trials" (2017), pp. 515–518.
74 Ibid. p. 516.
75 See Stolberg, Learned physicians (2022), p. 170, p. 277 and pp. 360–363; exam-
    ples are ÖNBW, Cod. 11251, "Experimenta quaedam brevia comparatu facilia
    vulgaria probata excerpta passim ab authoribus et secretis aliorum medicorum"
    (collected by Georg Handsch); ibid., foll. 296v–297r, on the entries in the *liber
    experimentorum* of a barber-surgeon; ibid., fol. 142r, "Experimenta et secreta
    Doct. Gerhardi Medici Archiducis Ferdinandi".
76 Falloppia, De bubone (1566), fol. 11v and fol. 14v.

77 Ibid., fol. 36r, "experimenta facta".
78 Ragland discusses Falloppia's use of "experimentum facere", in particular, in the context of the "experimentum" Falloppia claimed he had made on 1,100 men with the *linteolum* that was to protect them against the French disease. As we have seen, this *linteolum* was not a preservative, however. It was applied as a prophylactic remedy for several hours *post coitum* to destroy the morbid matter that might have remained in the skin in spite of washing. Ragland has found a somewhat different usage of "experimentum facere" in a passage in Mattioli's *Commentarii* (1554), pp. 566–567. According to Mattioli, one could "facere experimentum" and see that metals, except gold, could float on fluid mercury rather than sinking to the bottom. Clearly, this was still about proving a claim, however.
79 Falloppia, De bubone pestilenti (1566), fol. 14r.
80 Even Favaro (Gabrielle Falloppia (1928), p. 80), claimed that Falloppia dissected bodies "dopo avere sperimentato su di essi una tale azione sino alla dose letale."
81 Falloppia, De compositione (1570), fol. 13r; Falloppia, Libelli (1563), foll. 47v–48r.
82 ÖNBW, Cod. 11240, fol. 78r: "Verum omnes sudarunt nec insecuta sunt ulla symptomata qualia recenset Dioscurides."
83 Ibid., fol. 78r: "Dictum est de dr[achmis] ii, non tamen totum ebiberunt, quia maior pars residebat in fundo."
84 E.g. Falloppia, De aquis (1564), fol. 176r; see also the chapter on the French disease.
85 Falloppia, De tumoribus (1606), fol. 21v.
86 ÖNBW, Cod. 11210, fol. 192v; on the early modern notion of "imagination" as a cause of disease see Zaun, Watzke and Steigerwald, Imagination (2004).
87 Falloppia, De luxatis et fractis ossibus (1606), fol. 172r.
88 Falloppia, De bubone (1566), fol. 8v:

> Et quia ratio habet hanc malam conditionem, vt pro diuerso capite, varia videatur, et ille, qui plures habet authoritates, qui eloquentior est, magis persuadet. Ideo in rebus, in quibus est periculum vitae, primum inditium est ab experimento, et magis fidem habemus experimento quam rationi.

5 Medical practice

Contemporary sources described Falloppia unanimously as a much sought-after practitioner. High-ranking patients asked for his help and advice. Pope Julius III summoned him to Rome, in 1552, for the treatment of his sick brother Baldovino del Monte. Alfonso II d'Este, repeatedly consulted him, especially for his ailing sister Leonora. In July 1562, only a couple of months before his death, Falloppia spent ten days at the court in Ferrrara, at the request of Alfonso to take care of her.[1] Leonora, who had used the baths of Abano already in the preceding year, went back to Padua and Abano in the fall of 1562, together with her brother Luigi d'Este, to continue the treatment under Falloppia's supervision.[2]

Falloppia's medical practice, his disease theories, his explanatory models, and his diagnostic and therapeutic procedures in everyday practice have not been studied so far, undoubtedly due also to a scarcity of relevant sources. The student notes on his various lectures are not very revealing in this respect. *De caloribus* published in the *Opuscula* of 1566 briefly discusses the differences between elementary heat, natural or innate heat, and preternatural heat, their genesis and nature, and their place in the body in general and in the concoction of food and morbid matter and thus in the workings of remedies in the body, in particular.[3] The text is very brief, however. In his lectures on ulcers and tumors and especially in his detailed discussion of the French disease, where changes on the body's surface went hand in hand with pathological processes inside the body, his basic ideas about diseases and their treatment come to the fore more clearly. His lectures on simple and composite remedies, on plants, healing waters, metals, etc. offer some access to his ideas about the nature of "internal" diseases and the ways different remedies acted on them. But all this remains quite fragmentary.

Occasionally, the student notes on his lectures quote Falloppia not only referring to recipes and practical advice in general but also explaining with expressions such as "ego autem soleo" how he himself usually proceeded, sometimes adding "with success" ("cum successu") or "as I have experienced" ("expertus sum"). Sometimes he even mentioned concrete individual cases he had seen. His references to his personal observations are most

DOI: 10.4324/9781003242000-6

prominent in his account of the plague epidemic in Padua in 1555.[4] He had read the extant literature, he told his students, but a single day of that epidemic had taught him more than all his reading. As a warning to his students, he confessed his initial ignorance and what it led to. At the very beginning of the epidemic, he saw a patient and did not recognize the symptoms of plague. As it turned out, the man suffered from a pestilential fever and had a bubo at the neck. He had contracted the disease from infected cloth that had come from Istria and which carried the disease into three or four houses. If he and the others involved had recognized the bubo as such they could have prevented the epidemic, he felt. Drawing on his experience during the epidemic, he listed altogether nineteen symptoms of the plague, some of which could not be found in the books. The fever could not always be easily recognized because it was not always strong and some patients did not even feel it. The patients' own reports could not necessarily be trusted also for a different reason. Patients feared that they would be taken to a so-called "Lazarectum" if they were diagnosed with the plague. He had found several prostitutes, for example, who were laughing and seemed jolly and did not reveal that they had a bubo under the armpit. They were only found out because the husband or "moechus" (pimp) remarked that one of them had complained about a headache.[5] Among the typical symptoms of the plague, Falloppia listed the bubo, a patchy reddening of the face, stinking feces, and sweat but also a sign that nobody, he believed, had so far seen and described, namely many black blisters all over the body, filled with a yellow fluid.[6] As to therapy he warned against bloodletting, which most authorities recommended for the treatment of the plague. Experience showed that it was actually harmful. All the patients his colleague Frigimelica had subjected to bloodletting in 1555 died, while many of the others survived.[7]

Falloppia went into some detail in this specific case but he never taught *medicina practica* and we have nothing that comes even close to a systematic account of nosology and pathology from his mouth or quill. In what follows, I will therefore rely primarily on two types of sources: Falloppia's general precepts on the diagnosis, treatment, and prognosis of diseases, which have come down to us under the title *De modo consultandi*, and casuistic sources, above all the student notes on his oral pronouncements on individual cases on the occasion of the so-called *collegia*, the joint, public consultations by the Padua professors on individual patients.

De modo consultandi

De modo consultandi was first published in Falloppia's *Opera* in 1606; the more precise running title on the top of the pages was *De modo collegiandi*.[8] As he had already done with his lecture *De decoratione*, Falloppia seems to have established a Paduan tradition here.[9] From 1570s-Padua, manuscript notes by Johann Mattenberg on Girolamo Capivaccio's lecture *De modo collegiando seu consulendi* have survived.[10] Capivaccio's reflections on the topic later appeared in print in a much more detailed form.[11] Mattenberg also took brief notes on *De modo collegiandi* by Capivaccio's Paduan colleague Girolamo Mercuriale.[12] Falloppia, for his part, may have found some inspiration in his Pisa days from Giovanni Argenterio, who held the chair of practical medicine there. Argenterio published a detailed *De consultandi ratione liber* in 1551. According to Argenterio's dedication letter, he had intially written it for his own students and only later published a revised and no doubt more detailed version.[13]

Falloppia's *De modo consultandi* is strikingly concrete. He offered his students practical instructions, based on his personal experience. Falloppia started by pointing out two basic types of and occasions for a joint consultation of several physicians on the same case of illness. Sometimes high-ranking, noble patients, who always wanted to stand out from the rest, called several doctors for some minor illness. In this case, a joint consultation was useless and unworthy.[14] By contrast, in illnesses that were difficult to treat, the joint consultation of several doctors was useful and appropriate. The difficulty could be due to various reasons, in particular to the physical condition of the patient, the severity of the symptoms, and the duration of the disease. Sometimes the diagnosis already caused great problems, for example, in the case of apparently new diseases (he was probably thinking of the French disease), or when the actual location of the disease was difficult to identify, or when two different locations or organs in the body seemed to be affected. In such cases, bringing several experienced physicians together for a joint consultation was helpful or indeed indispensable. It also eased the pressure on the individual physician, Falloppia added, when the difficult disease (not to mention his possible ignorance) confused him and made him anxious.

In a consultation, it was first of all necessary to determine the nature of the disease. The patient and, if necessary and possible, the relatives, friends, and caregivers had to inform the physician about previous illnesses, the history of the current illness, and about possible causes and promoting influences, such as the diet and the patient's way of life. The physicians had to ask about the symptoms, such as pain, feelings of constriction, pulsations, contractions, unusual discharge, heat sensations, and the like. They also had to assess the physical condition and temperament of the patient, as revealed by what the patient and those around him reported and by physical signs, such as the abundance or lack of hair and its color. They needed to pay special attention to the central, vital organs and their respective temperaments, that is, the brain, the heart, and the liver. They had to explore the liver by manual examination.[15] Depending on the clinical picture, they also had to assess the state of other organs, such as the kidneys or the spleen.

Great caution was needed when the physician could not see the patient personally and had to rely on a written report. All too easily, Falloppia warned his students, the physician might attach too much importance to individual words and passages in the written account and draw wrong conclusions. Sometimes the epistolary account of a patient's illness barely revealed the name of the patient and the part of the body affected. Those who tried to make a diagnosis from such notes ("schedulis") were like those who claimed they could tell the illnesses of the country folks just by looking at their urine. This was unworthy of a learned physician.

When he knew enough about the patient and his or her illness, the physician could proceed to the actual consultation. In responses to the written request of a physician, one did not have to repeat in detail the *historia* of the illness in question. The addressee already knew it. The situation was different when several physicians came together to discuss a case at the bedside or in front of an audience during a *collegium*. Here, the physician who spoke first had to carefully describe the history of the disease. Since physicians often disagreed about the habitus and temperament of a patient or of individual parts of the body, the first physician speaking did well to point out the physical signs which led him to assume a certain temperament. Those after him, by contrast, could supplement these explanations, if necessary, or express their doubts on certain points but they should not, as some did, recount the patient's whole history again.

The detailed presentation of the patient's history and clinical picture was followed by the analysis. The physician had to determine the nature of the disease and the causes that triggered and sustained the disease, first the efficient cause, then the material cause, which as a rule was some kind of morbid matter. One had to go through the possible causes carefully and exclude those which played no role in the present case. If the patient suffered from several diseases at the same time, it was necessary to discuss whether they were related and whether one was possibly the cause of the other.

Sometimes very typical signs of a particular disease permitted a precise diagnosis or prognosis without further ado. On the other hand, if a physician was uncertain, he better left the judgment to those who spoke after him and, if his turn was later, he should not criticize those who had already proffered their judgment, as many did, driven by arrogance, in order to appear more knowledgeable than the others.[16] Falloppia also expressed his displeasure with physicians, however, who did not dare contradict their colleagues at all, even when it was necessary.

Prognosis posed particular challenges, in everyday medical practice, in dealing with the sick and their relatives. When a physician explained the pathological process inside the body, the patients and their families had no means to prove him wrong and if the treatment failed, it was still possible that the disease was incurable. By contrast, any ordinary farmer or craftsman could easily assess whether the physician's predictions on the future course of the illness eventually proved correct or not.[17] According to Falloppia, it was an all too familiar sight, for this reason, that physicians were reluctant to commit themselves on this point. Their prognosis sounded more like a Sibylline oracle. He thought he knew why: they feared for their reputation when they were caught making a false prediction. Some, he found, promised as a rule that the patient would recover, in order to make the patients and their relatives happy. Others predicted a difficult recovery or even warned of a possible fatal outcome, instead of the easy recovery they actually expected. In this way, they sought to make their successful treatment seem all the more praiseworthy. A pious physician ("pius medicus") did not do such things, he admonished his students.[18]

Falloppia's therapeutic recommendations reflected the accepted doctrine. One had to take the various external factors into consideration that could potentially promote the disease: the air, the food, the sleep, the affects of the soul, and the habits, in addition to the excretions and the natural and vital functions of the body.[19] The decisive factor for the successful treatment of most diseases, however, this was the therapeutic mantra in Padua at least since Giovanni Battista da Monte, was the elimination of the cause of the disease in the body.[20] Since Renaissance physicians attributed the vast majority of diseases to some morbid or peccant matter that was disseminated in the blood and/or settled in specific locations in the body, purgation was the central pillar of treatment in almost every illness. First, the physicians had to soften and mobilize the morbid matter and to relax the pathways in the body to facilitate its excretion. Then they had to evacuate it with purgatives which were known to exert a specific attraction on the morbid matter in question and, if necessary, prescribe bloodletting from a suitable vein. Once the morbid matter was evacuated, altering and strengthening medicines and local treatment could be applied, if necessary. These included, in particular, baths and drinking healing waters. Sometimes one also might have to recommend surgical intervention but surgery should be applied as mildly as possible, preferring caustics, for example, to the glowing cauterizing iron.

Medical cases

Only a few written consilia from Falloppia's pen have survived.[21] Falloppia never published case histories either, unlike his former colleague in Ferrara, Amatus Lusitanus, whose *Curationes* introduced a major new literary genre into learned medicine. A fair number of detailed oral diagnostic and therapeutic judgments statements by Falloppia have survived, however, in the form of student transcripts of the opinions he and other professors expressed on specific cases that were presented and publicly discussed in Padua in the so-called *collegia*. The *collegia* were a well-established and popular institution in Padua. They took place quite frequently and in different places – in the house of one of the professors, in private houses or even outdoors. In the mid-1550s, an unidentified student took notes on altogether almost fifty such *collegia* which he witnessed in Padua in the course of two years.[22] Some notes on *collegia* have also survived in manuscript.[23]

Others were published in print: thirty-two cases, in which the students documented Falloppia's judgment can be found in the 1587 edition of Vittore
Trincavella's *Consilia*.[24]

The *collegia* were an excellent didactic tool. They offered a unique opportunity for the acquisition of the knowledge, for learning the kind of reasoning and the skills the future physicians would need when they applied the
general theoretical disease concepts on the individual patient. They were
also a lesson in humility: the students experienced how even leading lumi-
naries could arrive at different conclusions about the same case. Last but
not least, they learned from the example of their professors how they could
later assert themselves when a distinguished patient called them together
with other physicians for a joint consultation.

The *collegia* followed a fairly strict order. First, the patient was presented
with his complaints and medical history. This was usually the task of the
lowest-ranking professor present. Sometimes he recited the essential aspects
on the basis of the written *historia morbi* of a patient, which the attending
physician sent to Padua, because the disease seemed particularly complex,
because the treatment so far did not have the desired effect, or because there
were several attending physicians who could not agree.[25] In other cases, the
patient must have been present in person, since the professors mentioned
that they had palpated his or her body with their own hands.[26]

Having presented the case, the first professor gave his verdict. He explained the nature of the disease and its external and internal causes and,
on that basis, gave his therapeutic advice. Then, the other professors, one
after the other, offered their judgment. Usually, three or four professors and
sometimes up to six professors participated in such a *collegium*. They rarely
were in complete disagreement but their judgments differed on some points.

The spectrum of clinical pictures that were discussed in the *collegia* was
broad. Some cases were complex and did not easily fit into a concrete, established disease category or clinical picture. A striking example was an elderly
patient who sought advice for his memory loss, impaired tongue movement,
a heavy head, hearing loss and ringing in the right ear, diminished vision in
the right eye, vertigo, catarrh, a bitter or pungent taste in the mouth, ulcers
on the palate, a tendency to nightmares, choking sensations when he bent
over, a tumor in the area of the stomach, joint pain, and syphilitic pustules
in the genital area and on the buttocks. Many of these symptoms could be
traced to rising fumes, the professors felt, especially to atrabiliary fumes but
this was not true for al all of them.[27] In many cases, the focus was on a more
or less clearly circumscribed clinical picture, however. Such cases were of
particular didactic value because the students could expect that they would
later have to deal with similar cases in their own practice.

The most frequent diagnosis in the surviving notes on *collegia* in which
Falloppia participated was melancholia. Other diagnoses included headache, dizziness, catarrh, loss of hearing and vision, spitting blood, dizziness,
asthma and heart tremors, colics, stone disease, and a hardening of the uterus.

There were also cases of patients with tumors, ulcers, and other pathological changes on the surface of the body. They belonged primarily to the realm of surgery but Falloppia and his colleagues attributed them to pathological changes inside the body that generated a morbid substance which accumulated near the skin or sought its way out via ulcers and fistulas.

Falloppia's typical diagnostic and therapeutic approach can best be illustrated by a concrete example. I will use the case of a seventy-year-old patient with a relatively clearly circumscribed clinical picture, which Falloppia, Bellocati, and Fracanzano discussed in a *collegium* on 20 March 1559.[28] Based on the written account of the patient's attending physicians, Falloppia first presented the history of the disease and the current symptoms. The patient complained of a burning pain in the urethra during urination, especially toward the end. The urine was sometimes pale ("albus") and sometimes yellow, and it contained different types of matter, such as small particles resembling bran and crushed eggshells, sand, and sometimes semen-like matter. The patient was slender and his "habitus" was cold and dry, as was typical for his age.

Having presented the case, Falloppia expressed his personal judgment. As outlined in his *De modo consultandi*, he started with an analysis of the possible causes. As in his other surviving contributions to *collegia*, Falloppia resorted to the prevailing doctrine of disease, which attributed almost all diseases to some impure, rotten, putrid, or else burnt morbid matter, a *materia peccans*, as Falloppia also called it. It moved into the individual parts of the body and often accumulated and hardened there but it could also continue to move through the body and ultimately be evacuated with the excretions. When it was slimy or viscous, it could also clog the veins and other pathways in the body or obstruct individual organs. Local accumulations of morbid matter could moreover generate a preternatural heat from putrefaction or release harmful vapors. These vapors rose upward in the body and sometimes liquefied in the head and from there flowed off again into the rest of the body as a "catarrh" – the term derives from the Greek words for "down" and "flow".

For the diagnosis and treatment of the disease, it was therefore crucial to identify, first of all, the nature and the preferred site of the morbid matter and, second, to identify the causes of its generation in order to be able to fight the disease at its roots. In the case of the seventy-year-old patient, Falloppia first discussed two possible causes of the burning pain. It could be due to an ulcer in the urinary tract. The urine which passed through it was acrid by nature. In the case of an ulcer, the attending physicians would have seen pus flowing out, however. Falloppia also rejected the possibility of kidney- or bladder-stones. The typical sediment and the characteristic pain were missing. The burning was more likely result from an alteration of the urine itself. Both the sand in the patient's urine and the small matter, the origin of which Falloppia suspected to be a salty mucus, gave the urine a certain pungency.

In a second step, which was again typical of the methodical approach of the Galenist physicians of that time, Falloppia asked for the source of the pungent pathogenic matter, which, he was convinced, together with the sensitive skin of the urethra, was responsible for the burning pain. As his tendency to develop diarrhea indicated, the patient had a weak stomach and therefore concocted the incoming food insufficiently. This resulted in raw, mucilaginous matter, which then was roasted and burned in the hot liver, as were other humors and fluids in the body, especially the yellow bile. The adust mucus and adust yellow bile passed to the kidneys and from there to the bladder, causing the burning. The brick-like and scaly eggshell-like matter in the urine arose from this adust matter, which solidified in the urinary ducts and bladder.

Having thus identified the (dual) nature of the morbid matter and its source, Falloppia proceeded to therapy. As in many other *collegia*, his therapeutic recommendations rested on two pillars above all, namely the evacuation of the morbid matter and the strengthening of the organs responsible for the generation and the reception of the disease matter. In order to cleanse the body of the morbid matter, he recommended various purgatives such as cassia, rhubarb, and senna. Bloodletting, which was done only exceptionally in old patients at the time, was not advisable but one could try to stimulate bleeding from the hemorrhoids. A sea voyage – here Falloppia presumably had seasickness in mind – would be an excellent means of promoting the vomiting of the mucous matter. Alternatively, an emetic could be given. Against the weakness of the stomach, which generated the salty, mucilaginous morbid matter, he recommended the waters of San Pietro in Montecatini or those of Villa near Lucca. They were not very hot and could at the same time temper and strengthen the liver and kidneys. In addition, he advised internal and external remedies to strengthen and moisten the urethra. As was typical for the everyday medical practice of Renaissance physicians,[29] dietetic recommendations – which loomed large in published medical writing of the time – played a marginal role. The student recorded only Falloppia's very brief and general recommendation on the temperate food patient should choose.[30]

In some cases, Falloppia, like his colleagues, resorted to additional, more specific explanatory models. Consumption, for example, was usually attributed to a sharp, catarrhal matter that flowed through the airways into the lungs, causing an ulcer. This was precisely the explanation Falloppia used in the case of an emaciated woman with fever and bloody sputum. His treatment consequently aimed at purging the interior of the chest with remedies such as hyssop or licorice juice, and at combating the formation of the salty, acrid matter by cooling the hot liver. In addition, the moist matter in the head itself had to be dried and, if that did not help, an artificial ulcer had to be created on the scalp, presumably in order to allow the morbid matter to exit this way.[31]

In a handwritten *consilium* for a man suffering from impotence, which he addressed directly to the patient, Falloppia refrained from a discussion of the possible causes but his suggestions for treatment implicitly reveal his interpretation of the disease process. Foods such as beans, truffles, and artichokes, which produced winds in the body, could have positive effects in this case; behind this clearly was the widely accepted idea that the penis of the impotent man was not sufficiently supplied with spirits that quite literally inflated it. Since this kind of diet was harmful in the long run, Falloppia also recommended morsels and a warming ointment for local treatment. It was also possible, he added, that the impotence was caused by coldness of the stomach, liver, and genitals themselves. In this case, the patient should consult a doctor and a visit to a thermal spring could help.[32]

The disease melancholia, in turn, – not to be confused with a mere melancholic temperament – was widely attributed at that time not to an excess of natural black bile but to a pathologically altered, namely burnt yellow or black bile (and sometimes also to black burnt blood).[33] In this sense, Falloppia attributed the sadness and anxiety of an aging patient and the melancholic "phantasma" that appeared to him in his sleep to a hot and dry liver and to the exhalations coming from the yellow and black bile that it produced abundantly and that ascended to the head.[34]

In one of the *collegia*, he also discussed the case of a young man who saw, in his right eye more than in his left, small particles that moved like mosquitoes – a symptom known to this day as "mouches volantes". According to Falloppia, the cornea and the lens were immaculate. He attributed the symptoms to vapors that moved constantly in front of the eyes, like flies that were flying around. These vapors came from a thickened, mucilaginous matter in the body that set these finer and airy parts free. His treatment therefore aimed mainly at ridding the head of this matter, with bloodletting and medicines that promoted the evacuation, via the saliva, in particular. The salty drip baths in San Pietro could help to dry the head and the fluid morbid matter. Cupping and cauterization of the arms were also to be considered.[35]

Repeatedly Falloppia built on his anatomical expertise to explain individual symptoms. For example, he attributed the pain of a patient with a tumor of the tongue to its consensus with the fourth and sixth pairs of the cranial nerves.[36] In other cases, he showed precise knowledge of anatomical structures to be even crucial for a correct understanding of the disease. A good example is the case of an eight- or nine-year-old boy who had repeatedly been afflicted by inflammation, dysentery, and catarrh since shortly after his birth.[37] More recently, he had thrown up or vomited blood and food mixed with black blood and now he had developed a persistent fever and dark stools. Falloppia located the origin of the evacuated blood in the liver, which produced an excessive amount of hot – and therefore particularly mobile – blood, more than the body could harbor. The fact that he ejected this blood mainly through the mouth indicated to Falloppia its origin in the deep thoracic veins. The boy also complained of pain that moved from his right upper abdomen to his neck, and the boy's parents told Falloppia that the veins of the neck swelled when the boy read aloud. As Falloppia explained, many veins extended from the liver to the throat but they were very fine. Falloppia saw the source of the blood elsewhere: it came from the veins called "coronariae" at the upper mouth of the stomach, that is, at the beginning of the esophagus. They had their origin in the *vena porta* and were close to the *vena cava*.[38]

Today, the rupture of enlarged and swollen veins or varices in the lower part of the esophagus is considered an important cause of vomiting blood, often associated with black, tarry stools. These veins swell when the blood flow through the liver is obstructed, for example, in liver cirrhosis, forcing the blood to take its route from the portal vein to the *vena cava* via small veins that are not designed for such large volumes of blood. They can therefore easily rupture, all the more so when there is a reflux of gastric acid into the esophagus. It is thus quite possible that Falloppia gave here the earliest surviving account of esophageal varices as a cause of hematemesis and bloody stools in chronic liver disease.

Anatomical knowledge, combined with manual skills, was particularly helpful in surgical cases. The case of a patient with an anal fistula, which Falloppia discussed in a *collegium* with Bellocati, Fracanzano, and Trincavella, illustrates this nicely. Falloppia examined the patient carefully. He diagnosed a rather large abscess in the anal region, extending mainly to the left side, from which half a pound of pus had oozed. Using a probe, he was able to penetrate four to five finger widths deep, and he found two openings to the outside, one of them visible at first glance, the other one seemingly closed. However, as could be shown by injections – presumably he meant the injection of fluid into the other opening – the abscess was also connected to the body surface via this apparently "blind" opening. The edges touched but were not fused. He could not reach this opening from the abscess on the right with a probe or with his finger but he suspected that this was due only to a fleshy mass that was in the way and that he could feel. He concluded that the abscess or fistula did not penetrate the anal sphincter entirely because in that case, the patient would be incontinent. He believed the middle part of the sphincter to be affected, however, because when he advanced his finger, the sphincter did not contract hardly. His judgment thus rested largely on his anatomical knowledge of the precise site and extension of the anal sphincter.

Falloppia did not recommend a surgical treatment of the fistula in the narrower sense. The condition had existed for too long and was subject to a constant influx of fluids and excrements. Its precise extension was unclear and the anal sphincter must by no means be damaged. He therefore recommended a "curatio palliativa". One should keep the fistula open, with some cloth or, better still, with a silver cannula, at least temporarily, if much matter was draining. In addition, the influx of morbid matter had to be stemmed, as far as possible, given the weakness of the parts concerned, and the formation of new excremental matter had to be prevented by an appropriate diet. The patient had to avoid food, in particular, that was difficult to concoct and, because it remained somewhat raw, promoted the formation of mucus. He also recommended purging the whole body and regular bloodletting on the arm.

To sum up: Falloppia's diagnostic and therapeutic approach was quite conventional. Even in recent work on early modern medicine, the view persists that the Galenic physicians of the sixteenth and early seventeenth centuries generally attributed diseases to an "imbalance" of the four natural humors in the body, to an excess or deficiency of blood, yellow or black bile, or phlegm, or to an imbalance of the primary qualities – warm, cold, hot, and dry – associated with each humor in pairs. As I have shown elsewhere,[39] the interpretation of diseases as a result of a disturbed balance of the humors or qualities in the body was largely irrelevant in everyday medical practice, however. In literally hundreds of medical *consilia* and *observationes* that have come down to us from the sixteenth century, cases in which the physicians attributed diseases to an imbalance of the humors or qualities in the body must be searched for like a needle in a haystack. At most, the physicians suspected a local *intemperies* in a specific organ. In this sense, a cold stomach could coexist with a hot liver in one and the same patient. The crucial pathological factor, however, was ultimately the formation of morbid matter, which might be slimy and raw in the case of a cold stomach and sharp and biting, by contrast, when the liver was too hot.[40] This was exactly the explanatory model Falloppia applied in the *collegia*.

Two somewhat special characteristics of Falloppia's approach catch the eye, however. One is Falloppia's great confidence in the beneficial effects of healing waters. They were widely used in medical practice at that time but Falloppia seems to have perceived them virtually as a panacea. There was hardly a patient for whom he did not recommend, among other things, visiting a certain thermal spring or at least drinking its waters. Presumably, he also saw this as a good opportunity to show off his personal expertise in this field. After all, he devoted a series of lectures to healing waters and presented the results of his chemical analysis of a range of thermal springs as well as his personal observations on their effects in different kinds of diseases.

A second, more fundamental feature was the extent to which Falloppia drew on detailed anatomical knowledge, when he could, to explain a patient's disease process or symptoms. His colleagues, too, sometimes referred to anatomy but in this field, Falloppia was able to bring his superior anatomical expertise to bear, underscoring once more, in turn, the importance of anatomical knowledge for medical practice.

Notes

1 Favaro, Gabrielle Falloppia (1928), p. 112.
2 Ibid., pp. 112–113.

3 Falloppia, De caloribus (1566); the text is very similar in content to the *Digreßio de concoctione* in the 1563 edition of Falloppia, De tumoribus (1563), foll. 57v–62v and may well be simply an excerpt from student notes on his lectures on this topic.

4 Falloppia, De bubone (1566).

5 Ibid., fol. 4v.

6 Ibid., foll. 3v–4v.

7 Ibid., foll. 8v–11r.

8 Falloppia, De modo consultandi, in idem, Opera (1606), vol. 3, foll. 192r–194r; according to the publishers, the brothers De Franciscis, they owed the new texts in this edition to two manuscript volumes by Andrea Marcolino, which he had given to their father. This makes Marcolino the most likely writer; the publishers of the Frankfurt edition reprinted the text, in the same year, on pp. 94–98 of *Appendix* to their 1600 edition.

9 Bassiano Landi also seems to have offered instructions of this kind. The notes on his *Ratio habenda consultationum* in Biblioteca Nazionale di Napoli, Ms. V H 203, fol. 23r–v, are very brief, however, more a list of bullet points.

10 Forschungsbibliothek Gotha, Chart. A 629, foll. 53r–54r.

11 Capivaccio, De medica consultandi ratione (1603).

12 Forschungsbibliothek Gotha, Chart. A 629, foll. 55r–v.

13 Argenterio, De consultationibus (1551); inside the book, the treatise carries the title "De consultandi ratione liber".

14 Falloppia, De modo consultandi (1606), fol. 192r.

15 Ibid., fol. 192v, "tactuque praecipue tendandum hoc viscus est".

16 Ibid., fol. 193v.

17 See Stolberg, Doctor-patient relationship (2021) for a detailed analysis.

18 Falloppia, De modo consultandi (1606), foll. 193r–v.

19 Ibid., fol. 193v.

20 Stolberg, Learned physicans (2022), esp. pp. 181–182.

21 Staatsbibliothek Berlin, Slg. Darmstaedter 3c 1550, letter from Gabrielle Falloppia to Giovanni Francesco Canani, Padua, 1 April 1561, on the case of a woman who suffered from headaches, dizziness and numbness in her fingers; Biblioteca palatina di Parma, autografi. fondo di Lucca, carteggio Beccadelli, epistolary consultation for the Archbishop of Ragusa, L. Boccadelli, 1 September 1562, ed. in Di Pietro, Epistolario (1970), p. 76; Biblioteca vaticana, Reg. lat. 1297, pp. 191–193, copy of an epistolary consilium by Falloppia for an unidentified man who complained of impotence.

22 Herzog August Bibliothek Wolfenbüttel, Cod. Guelf. 20.22 Aug. 4°, foll. 1v–130v, "Quaedam excerpta a me ex collegijs medicorum patauinorum Anno 55 et 56"; notes on Falloppia's contributions are on foll. 1v–2v, foll. 5v–7r, foll. 13r–14r, foll. 15v–16v, foll. 24–25r, foll. 26v–27r, foll. 29v–30v, foll. 46v–47r, and foll. 81v–83r.

23 Universitätsbibliothek Erlangen, Ms. 910, foll. 50r–51r and foll. 150r–152v, notes bei Johannes Brünsterer on two *collegia* with Falloppia, one about a woman with breast cancer and one about a man with vertigo, cramps, and other complaints which the physicians attributed to the brain and the nerves. In both manuscripts, there are notes on further *collegia*, which do not indicate the name of the professor who presented the case and was the first to give his opinion. In some cases, at least, this may have been Falloppia as well. Biblioteca comunale di Siena, Misc. XVI, C IX 32, foll. 1r–15v, notes by an unidentified student on three *collegia*, in which Falloppia participated in March and April 1559.

24 Trincavella, Consilia (1587): book 1, n° 7, 15, 20, 53, 68, 72, 75, and 76; book 2, n° 7, 9 and 14; book 3: n° 8, 10, 22, 45–48, 53, 56, 61, 69, 72, 78, 85, 100, 106, 107, 111, 114, and 119.

25 In one case, Falloppia, Bellocati, and Trincavella were explicitly asked for their advice "because of the discord among other physicians" ("propter discordiam aliorum medicorum"; Trincavella, Consilia (1587), col. 453).
26 For example, Leonicus on a female patient with stomach troubles in Trincavella, Consilia (1587), col. 333: "Ego tactu percepi, in contento, non autem continente pilae similem tumorem, quodque magis versus dorsum vergat."
27 Trincavella, Consilia (1587), coll. 325–327.
28 Biblioteca comunale di Siena, Misc. XVI, C IX 32, foll. 1r–3v.
29 Stolberg, Learned physicians (2022), pp. 124–129.
30 Biblioteca comunale di Siena, Misc. XVI, C IX 32, foll. 3r: "Victus mediocris esset, non calidus, nec attenuans."
31 Trincavella, Consilia (1587), coll. 262–263; in this case, the dietetic recommendations were more elaborate.
32 Biblioteca Vaticana, Rome, Reg. lat. 1297, pp. 191–193.
33 Cf. Stolberg, Learned medicine (2021), pp. 282–289.
34 Trincavella, Consilia (1587), coll. 325–327.
35 Ibid., coll. 196–197; Bellocati, by contrast, warned of a danger of epilepsy and objected against bloodletting.
36 Ibid., coll. 669–670.
37 Ibid., coll. 217–219.
38 Ibid.: "Mihi tamen videtur ab ore ventriculi superioris [sic!] manasse, quae etiam coronariae dicuntur, ortae a vena portae, venae cauae admodum propinqua [sic!]."
39 See my detailed analysis in Stolberg, Learned medicine (2022), pp. 155–315.
40 Ibid., pp. 140–145.

6   Last years

Leaving Padua?

By the mid-1550s, Falloppia had found his place in Padua, it would seem. He was a renowned teacher who attracted students from all over Europe. He ran a flourishing medical and surgical practice. Princes and even the Pope asked for his medical advice. But apparently he was not happy with his situation. He was looking for a change. With the support of Ulisse Aldrovandi, who was well connected in Bologna, he sought to obtain a lectureship at the University of Bologna.

The first signs of Falloppia's efforts in this direction are documented for 1557. In June of that year, Lelio Ruini recommended that Falloppia be appointed in Bologna. Falloppia, Ruini pointed out, enjoyed an excellent reputation. He was very popular among the students and was likely to attract numerous students to Bologna.[1] For the time being, the matter seems to have fizzled out but Falloppia must have continued his efforts. In two letters to Aldrovandi, in November 1558 and in January 1559, he vaguely referred to his "case" and asked Aldrovandi for his support.[2] Apparently he was successful. In February 1559, Camillo Canonici embarked on negotiations with Falloppia on behalf of the Bolognese Senate, the *Quaranta*. According to Canonici, Falloppia demanded a contract for no more than six years and 250 scudi salary for the first two years and 300 scudi for the remaining four.[3] For unknown reasons, the negotiations failed, however.

In January 1561, Falloppia returned to the matter again. He expressed his regret for Aldrovandi, who had given up his lectureship in the arts for the one on the simples, and explained that he himself desired to be freed of his teaching obligations in anatomy.[4] This time, things moved forward. Around Easter 1561, as we learn from a letter Falloppia wrote to Aldrovandi in October 1561, Canonici promised him an appointment, once he was free to leave Padua. As Falloppia explained to Aldrovandi, he would gladly come, though not to lecture on surgery, as he had said to Canonici. He hoped to obtain the aging Antonio Maria Betti's ordinary professorship in practical medicine instead – Betti died in 1562, in fact – or the chair of theoretical medicine which had become vacant with Benedetto Vittore's death.[5]

DOI: 10.4324/9781003242000-7

The *Quaranta* instructed their secretary Galeazzo Zambeccari to negotiate the terms of the contract with Falloppia.[6] Falloppia's pupil Giovanni Battista Carcano Leone (1536–1606)[7] later claimed that Falloppia at that time had already suggested that Carcano take over his lectureship in Padua on anatomy and surgery, which he wanted to give up for health reasons (apparently he did not tell Carcano about his plans to go to Bologna).[8] Falloppia's hopes were never to materialize, however. His contract at University of Padua ran until 1563 and the Venetian authorities did not let him go, it seems. He could have – and very well might have – come to Bologna in the fall of 1563 with the start of the new academic year but he died before he could embark on this new stage of his career.

The reasons why Falloppia wanted to leave Padua are not entirely clear.[9] Practical considerations may have played a role: Bologna was much closer than Padua to his native Modena, where he stilled owned a house. In his letters to Aldrovandi, Falloppia stressed above all his wish to get rid of teaching obligations in anatomy. Anatomical studies were suitable for a younger age ("età virile"), he maintained. In later years, it was time to devote oneself more to theoretical questions, to "speculation". He wished to focus now on medicine only.[10] Performing anatomies in the cold winter undoubtedly entailed certain hardships. The public anatomical demonstration only lasted for a few weeks every year, however. And there is little to suggest that his interest in anatomical work as such diminished. After all, he wrote and published the *Observationes anatomicae* in those years and announced an even much more voluminous, comprehensive work on human anatomy with numerous illustrations.

A more plausible explanation for Falloppia's wish to move to Bologna, it seems to me, is that he was unsatisfied with his position and status among his colleagues. Falloppia's teaching was a major reason why numerous students and aspiring doctors came to Padua. For their part, the *Riformatori* and the Senate in Venice repeatedly let it be known that they were well aware of Falloppia's significant contribution to the reputation of the University. However, the teaching of anatomy and surgery came with a markedly lower position within the professorial hierarchy than that of theoretical and practical medicine. This found its tangible expression in the salaries. Even though he combined the lectureship in surgery and anatomy with the one on the simples, Falloppia was awarded an annual salary of 200 fl. only.[11] The professors of theoretical medicine, Bassiano Landi and Oddo degli Oddi, by contrast, received more than double the amount, 300 and 500 fl., respectively, and the remuneration of the professors of practical medicine, Antonio Fracanzano and Vittore Trincavella, was even more generous, with 350 and 950 fl;[12] and Trincavella saw his salary further increased to 1,200 fl. in 1554.[13]

Tensions and conflicts

Falloppia's subordinate and humiliating status within the professorial hierarchy was made plain for everyone to see during the *collegia*, the joint consultations of various professors on individual patients in front of the students for which Padua was famous. As student notes on the *collegia* in which Falloppia participated show, Falloppia was forced again and again, in front of the assembled students, into the role which was usually assigned to the lowest ranking and least experienced participant. He had to present the case in question and then give his opinion on it. After him, the professors of the more prestigious chairs had their say, one after the other. In the more than thirty recorded *collegia* in which he took part, it was almost always Falloppia who had to present the case and who was the first to offer his judgment. I have only found three exceptions: once Falloppia was preceded by Alessandro Massaria (1510–1598), who was not even a lecturer, and on another occasion by an unnamed physician, to whom the student notes referred as "N." and who was likewise quite possibly not a member of the university.[14] Once only, in 1559, Alvise Bellocati, another professor, spoke before Falloppia did – and in another *collegium*, the sequence was reversed again.[15]

On some occasions, four or five other professors would speak after Falloppia. So it was Falloppia who, in April 1552, had to present the case of a woman suffering from breast cancer and was the first to give his verdict, followed by Leonicus, Crassus, Frigimelica, and finally Trincavella.[16] Like a novice, Falloppia had to expect that the professors who spoke after him would contradict him – which they did again and again, if only on matters of minor detail. If Falloppia recommended cauterization, for example, a colleague might declare that it was dangerous in that specific case. The incumbent of the most prestigious chair had the last word. In many of the documented collegia in which Falloppia participated, this was Vittore Trincavella, whose son published the corresponding student notes. Born in 1498, Trincavella was a well-known and respected practitioner but toward the end of Falloppia's time in Padua, he was already considered an old man by the standards of the time. As Georg Purkircher reported shortly after Falloppia's death, his students were in fact dissatisfied with his teaching and laughed at him because his weak memory often failed him.[17] In short, from a didactic perspective, the *collegia* were an excellent means of introducing students into the art of diagnostic and therapeutic judgment and disputation with medical colleagues. But the differences of opinion and controversies that inevitably arose about the nature and treatment of individual cases were likely to foster tensions and conflicts among the professors – and they offered ample opportunities to the senior professors to put the lower-ranking ones in their place.

Sometimes the professors carried out rivalries also in the absence of their opponents in their lectures. They could count on the students to act as intermediaries, who would let the respective opponent know what had been said against them. In this arena, Falloppia with his numerous students was in a better position than during the *collegia*. As the lecture notes of his students show, he was no stranger to controversy. In his lecture on injuries, he went to particular lengths to reject the opinion of "some of the physicians at this university", who with "futile arguments" and without being able to refute his own view, opposed his (quite unconventional) understanding of the "solutio continui".[18] The "solutio continui" – this calls for an explanation – was the basic category of disease under which injuries (as well as ulcers and other skin lesions) were grouped. Falloppia opposed the common view, held by his opponents, that the "solutio continui" concerned both the *partes similares*, that is, roughly the tissues in modern terms, and the *partes organicae* or *instrumentales*, that is, roughly the organs; "organ" and "instrument" originally both meant "tool". In his opinion, and here he implicitly applied Galen's definition of disease as an impairment of function, the *solutio continui* was a disease of the *partes instrumentales* or the organs only. A "solutio continui" led to a partial or complete loss of function. Falloppia declared the claim implausible that a "solutio continui", that is, some local damage, to a *pars similaris* decisively affected the specific function of the diseased part of the body. Admittedly, the substance of bones and nerves, for example, counted among the *partes similares* but when a bone was fractured or a nerve injured that specific bone or nerve could no longer perform its own ("proprium") specific action, because that action depended on the form, size, location, etc. that allowed it to fulfill function. As Falloppia related their views, his opponents supported their position, among others, by maintaining that when flesh was generated and filled the defect, there was an excrement, that is, pus or *sanies*. This excrement could only result from a *pars similaris*, which proved in their eyes that the wound was a disease of the *pars similaris*. Falloppia first of all denied that there was always *sanies* in wounds. Some wounds healed within a couple of days, without any *sanies*. Moreover, the *sanies* that could in fact often be observed did not result from a damage to the faculty of the *pars similaris* due to the wound as his opponents claimed. It was not caused by the *solutio continui* at all but by an *intemperies* which resulted from the exposition of the unprotected wounded flesh to the outside.

This was not just a matter of differences of opinion. Falloppia combined his refutation with a massive attack on the surgical competence of his opponents. They presented themselves in the *collegia* and on other occasions as highly skilled in surgery and as if they were well versed in this art but their poor judgment on this issue proved the opposite.[19] Falloppia initially referred to physicians at Padua University in the plural but, as the argument progressed, he repeatedly spoke more concretely of his "foe" ("adversarius"), whose various arguments, as they were brought to his knowledge by the students ("vt mihi relatum est tale"), he contradicted at length. The student notes suggest that Falloppia did not explicitly mention the name of this "adversarius" but his audience undoubtedly knew whom he had in mind. In the printed edition of Falloppia's writings, a note in the margin, presumably by Marcolino, spelled the name out. It was Trincavella, who had criticized Falloppia's point of view and whose opinion Falloppia now harshly rebuked – asserting his own, superior authority in surgical matters, not only as a practitioner but also when it came to theoretical argument.

Falloppia was in a particularly strong position when it came to anatomical matters. In discussions about individual cases and controversies about theoretical questions, there was no clear "right" or "wrong", apart from controversies about the existence of relevant passages in the works of the ancients. The persuasiveness and plausibility of the arguments – combined with the reputation of the person who offered them – would ultimately tip the scales in favor of one professor or another. In controversies on anatomical matters, by contrast, the opinions and alleged findings of an opponent could be tested and disproven beyond doubt by examining and dissecting corpses. When the professor of *medicina theorica* Bassiano Landi explained to his students with Galen that "there is no experienced anatomist who does not recognize that the optical nerves are hollow",[20] and Falloppia, on his part, showed his students in the dissection room that the optical nerves had no duct,[21] he inevitably damaged Landi's authority. Landi had been teaching anatomy for many years[22] and had even published an extensive, erudite anatomical textbook, in 1542, with the same Basel printer, Oporinus, who printed Vesalius' famous *Fabrica* the following year.[23] It was all the more with a sense of satisfaction, it would seem, that, a couple of years later, Falloppia reported to Ulisse Aldrovandi that the students were beating down on Landi shouting with great vehemence that they wanted someone else to teach theoretical medicine.[24]

On the occasion of public anatomical demonstrations, such controversies were sometimes fought out in front of the whole student body. Surviving manuscript student notes indicate that the professors of practical and theoretical medicine were sometimes present during Falloppia's public anatomical demonstrations and disputed with him. So, according to Handsch's notes, Falloppia once explained to his students that the claim that virgins had a hymen in the vagina was a fable. "But D. Frankenzanus said", Handsch's entry continues, "that while no such membrane was found in healthy persons, in very rare cases the vagina was preternaturally closed by this membrane, and the women concerned could not menstruate and conceive".[25] When Falloppia explained the vessels and nerves of the arms and legs, he stated that bloodletting could be performed on the *vena poplitea* (in the backside of the knee) only very rarely and with great difficulty. On this, three professors who were present at the event gave their comments. Fracanzano confirmed that even Vesalius could not find the vein in a woman. Oddo degli Oddi (1478–1558), professor of practical medicine, said that he had once seen it in a skinny woman, and Bellocati claimed that he had once bled a Greek man from this vein.[26]

If, in this case, the professors still agreed to some extent, Oddo degli Oddi, on another occasion, publicly and openly challenged Falloppia's anatomical expertise. The Helmstedt Anonymus documented the controversy in detail.[27] Oddo first expressed doubts about what Falloppia had said in a lecture, when he demonstrated the *vena azygos*, the large vein in the thorax (which, according to modern understanding, branches out into the entire thoracic cavity). According to Falloppia, it was correct to perform bloodletting on a vein on the right side in patients with pleurisy. Oddo cited a passage from the Hippocratic *De ratione victus in morbis acutis* against Falloppia, which advised the physician to decide according to the precise location. If the upper part of the thorax was affected, bloodletting was appropriate. If it was the lower part, one had to give medicines. Falloppia responded, citing Galen in turn, that *De ratione* was not by Hippocrates. According to the notes of the Helmstedt Anonymus, Falloppia actually wanted to elaborate but the students' noise prevented him from doing so.[28] At the end of the demonstration, Falloppia showed the students the *vena azygos*, which nourished the ribs as well as the lower parts, thus offering anatomical evidence that there was no rationale for treating pleurisy in the upper part differently from that in the lower parts.[29]

Oddo also raised questions about the origin of the veins in the body. Falloppia had stated, in line with traditional Galenic doctrine, that all veins had their origin in the liver. But this, according to Oddo, was wrong, for the *venae arteriales* (the *arteriae pulmonales* of modern nomenclature), which carried the blood from the right ventricle to the lungs, had their origin in the heart, not in the liver. Falloppia, according to the student notes, did not answer, because of the "familiar shouting" of the students.[30] Presumably, they understood that Oddo's attack was unjustified. Falloppia had explicitly stated that the *vena arterialis* owed its name only to its function – it brought blood from the right ventricle to the lungs – but that with regard to its build, it was an artery.[31]

Rivalries among professors were not uncommon in the universities of that time. In Padua, they were encouraged and promoted by the fact that, as a rule, two professors held their lectures at the same time of the day, in direct competition with each other. In Falloppia's time, Oddo degli Oddi, for example, was the *concurrens* of Bassiano Landi, in theoretical medicine, and Fracanzano that of Trincavella in practical medicine. When Joachim Curaeus came to Padua in 1557, he found that Fracanzano managed to attract the students with his sweet-talk and was more popular than the highly learned Trincavella.[32] According to Bonifaz Zwinger, the decisive reason why even the famous Girolamo Mercuriale (1530–1606) left Padua and accepted a professorship in Bologna – albeit a very well-paid one – was that his rival Capivaccio made life difficult for him, inciting others, wherever possible, to disrupt Mercuriale's lectures.[33]

The anatomical professorship was an exception. Falloppia had no direct competitor who lectured at the same time as he did. Apparently, however, the rivalries with Falloppia, the very successful and famous but lower-ranking professor of anatomy, were all the more pronounced. The other professors of medicine sought to severely restrict Falloppia's anatomical teaching, claiming that it exceeded the boundaries set by the university statutes. As the student notes of Handsch and the Helmstedt Anonymus from those years show, Falloppia taught anatomy in its full extent. He not only dissected the cadaver with his own hands but introduced his anatomical

demonstrations with detailed explanations in the style of a conventional lecture. In Padua, however, as in other universities of the time, anatomy was
traditionally taught with divided roles. A professor of lower rank read the
relevant passages from Mondino da Luzzi's classic anatomical textbook.
The chair, a professor of practical or theoretical medicine, commented on
these passages and showed the relevant parts on the cadaver, which a surgeon, the *sector* or *incisor*, laid open. Vesalius had temporarily combined the
functions of lecturer, demonstrator, and dissector in his person. However,
at a public anatomy in Bologna in 1540, he, too, had to accept that Matteo
Corti gave the actual lecture on Mondino's standard anatomical text, which
Vesalius subsequently contradicted and corrected in front of the assembled
studentship.[34] In Padua, in the years before Falloppia's arrival, anatomy was
also taught in the old style again. Still in 1550/51, Georg Handsch made detailed notes on a public anatomy where the roles were distributed according
to the professors' standing. At first, Handsch mentioned only Antonio Fracanzano as "legente et demonstrante" but later he added that the anatomy
had lasted for fourteen days, with Andrea Ap(p)ellato "legente" and Alessandro Veronese doing the actual manual work of dissecting the corpse.[35]

After his arrival in Padua, Falloppia reunited the three roles again. In
his appointment decree, his competences were not precisely defined. As
far as the simples and surgery were concerned, he was explicitly called to
lecture: he was to interpret the relevant books, that is, comment on them.
With regard to anatomy, by contrast, it was only stated that he was to offer
what anatomists usually offered.[36] Only when Falloppia was confirmed in
his post a year later, after his return from Rome, he was expressly told to
lecture on surgery and the simples "and to cut, read and show the anatomy,
as he has done last year".[37]

Those in charge in Venice thus knew that Falloppia had not limited himself to dissecting the corpses and they wanted him to continue teaching
anatomy in its entirety. But Falloppia's anatomical lecturing aroused the
resentment of his colleagues it seems. Bassiano Landi, in particular, did not
stop lecturing on anatomy after Falloppia's arrival in 1551. In December of
1553, the *Riformatori* in Venice wrote to those in charge at the University of
Padua that Landi had complained that the rector of the arts faculty, without
even listening to him, had reduced his salary because he had lectured on
anatomy.[38] Presumably, the rector felt that Landi had arrogated to himself
to lecture on a subject that was Falloppia's. We do not know whether Falloppia filed a complaint but Landi surely must have suspected that Falloppia
was behind this. The *Riformatori* agreed with Landi and demanded that he
be paid his full salary.[39]

The following year, the conflicts broke out even more massively, this time
probably instigated by Landi or other professors. In December 1554, the
vice-rector and the councilors of the arts faculty answered a letter of the
Riformatori, which unfortunately does not seem to have survived. Apparently, the *Riformatori* had asked the arts faculty to comply fully with the

statutes of the university and the faculty had therefore decided that in the current (academic) year, the first professor of practical medicine, Vittore Trincavella, would give the anatomical lecture, while Gabrielle Falloppia would only dissect and demonstrate, as the statutes stipulated. With this letter, the arts faculty asked the *Riformatori* to confirm this choice.[40]

Apparently, the faculty was instructed, in response, to divide the roles in the traditional manner described above: Apellato, professor of *medicina practica*, was to read the relevant passages from the textbook of Mondino da' Luzzi. Trincavella was to explain and show as the presiding professor. Falloppia was to dissect. But in January 1555, the university had to report to the Venetian authorities that this plan had failed miserably. Apellato did not get to read Mondino's text, as Trincavella alone lectured on anatomy instead. According to the vice-rector and the councilors of the arts faculty, Trincavella's first lecture, in the university building, still went smoothly. The second, however, was cut short prematurely and when he tried to teach during the anatomical demonstration, the students showed themselves very dissatisfied. The next time, Trincavella went about lecturing, they forced him to stop, shouting loudly "We want Falloppia!" Falloppia, the report continues, tried to quiet them down. They would be satisfied with what Trincavella had to say. But the students continued to chant: "We want Falloppia!" The following day, Trincavella tried again but, because of the noise, no word could be understood and he could not deliver his lecture. In the afternoon, the students categorically refused to let Trincavella continue lecturing. The anatomical demonstration was discontinued.[41]

The arts faculty underlined the great damage which the university suffered from these events. Many students, they explained, had stayed in Padua specifically for the anatomical demonstration, others had even come to Padua from other universities for that reason. The arts faculty therefore urged the *Riformatori* to allow them to offer the anatomical teaching the way it had been done before, especially since everyone was calling for Falloppia. The weather still allowed to hold an anatomical demonstration and the *Podestà* of Padua had a criminal on hand for execution whom he had promised to leave to the anatomists for dissection.[42] It must have been an extremely humiliating experience for the aging Trincavella. And it was a triumph for Falloppia. With good reason, the arts faculty emphasized that Falloppia had sought to calm the students. No doubt they wanted to divert the suspicion that Falloppia might have instigated these riots.

So far, I have found no evidence that the *Riformatori* actually revoked their decision, which was, after all, in line with the statutes. It is therefore quite possible that Falloppia was no longer allowed to give formal lectures on anatomy or that the conflict remained at least unresolved. In the year after the "riots", when lectures began again in November 1555 after the severe plague epidemic in Padua, the students waited in vain for a public anatomical demonstration. The fault was the mutual envy and dislike among the professors, Georg Keller reported at the end of January 1556. One of them, Keller complained, clearly alluding to Falloppia, claimed to be sick. Trincavella, on his part, pretended he could not get cadavers when Keller, on behalf of the *Natio germanica*, sought to convince him to teach anatomy, since no one else was in the way now.[43] In the following winter of 1556/57, Falloppia asked the *Riformatori* for assistance in obtaining a cadaver and promised a nice anatomy, also on a monkey and a bear.[44] It seems that he did hold at least one anatomical demonstration, because he complained of a cold that he contracted,[45] but it is unclear who did the lecturing. The warnings of the *Riformatori* in the following year suggest a continuing unrest among the students. Twice within eight days, the rectors of the university were admonished in December 1557 that they should make the necessary arrangements to avoid tumults, ensure orderly proceedings, and punish the disobedient students if necessary.[46]

Gabrielle Falloppia and Melchior Wieland

Falloppia never married and, for all we know, had no children. He lost his own father as a child and his mother died around 1550. His most important companion in his Padua years was, for all we know, Melchior Wieland or Melchiorre Guilandino, as the Italians called him.[47] Wieland had an unsteady life behind him when he came from Rome to Padua in the early 1550s. Little is known about his childhood and youth in Königsberg but he seems to have enjoyed a solid education in the arts. As he himself later recounted, he went to Calabria as a young man and lived partly by selling medicinal plants. Perhaps he came to the attention of Marino Cavalli, the Venetian envoy in Rome,[48] who, knowing of their common botanical interests, brought him into contact with Falloppia. It is also possible that Falloppia made Wieland's acquaintance in Rome, in 1552, when he spent some time there to treat the Pope's sick brother. At any rate, Falloppia took him into his home and the two lived together until Falloppia's death.[49]

In 1558, Wieland embarked on an extended research trip to the Orient. In February 1558, he described his plans: he wanted to travel to Constantinople with Marino Cavalli, who was now the new Venetian envoy to the Turkish ruler, and from there to Egypt, Arabia, Mesopotamia, Armenia, Persia, and "countless" other regions as far as the Moluccas.[50] According to Mattioli, Falloppia gave Wieland seventy gold ducats for the trip.[51] From his journey, he wrote enthusiastically to Aldrovandi about how much he was learning and discovering.[52] Then, however, he was captured by North African pirates. Falloppia managed to free Wieland from captivity, paying 200 gold pieces, the story goes.[53] According to Favaro, Falloppia even traveled personally to Greece to free Wieland.[54] Still after Falloppia's death at any rate, Wieland thanked him with effusive words for his generosity in freeing him from the chains of the "Numidians and Moors".[55]

Falloppia probably also played a decisive role in Melchior Wieland's appointment as director of the botanical garden in September 1561, as the successor of Luigi Squalermo (1512–1570) aka Anguillara.[56] Falloppia and Anguillara were apparently not on particularly good terms. Perhaps they came into conflict with each other over the botanical garden. With undisguised glee, Falloppia reported to Aldrovandi in 1554 how Anguillara had disgraced himself in Milan, as Falloppia had learned from his Milanese correspondents. Anguillara pretended that he was a lecturer at the University in Padua – which was not true – and his lie was discovered.[57] In two letters to Duke Alfonso II d'Este in 1560, Falloppia nevertheless recommended his "caro compadre" in the highest terms for his excellent knowledge of plants, metals, and minerals and as someone who would also be able to do alchemy if it were not "a vain art". He suggested to the Duke, who himself liked to deal with "practical natural philosophy", to bring Anguillara to Ferrara.[58] Anguillara was called to Ferrara, which paved the way for Wieland who directed the botanical garden until his death in 1589.

The controversy between Mattioli and Wieland

The Renaissance was a time of heated scholarly controversies. Falloppia himself seems to have been inclined more toward the irenic side. Even his critique in the *Observationes anatomicae* of Vesalius' numerous errors and gaps did not give rise to a serious conflict, certainly thanks, above all, to the respectful tone that Falloppia had used. He did not remain untouched, however, by the massive conflicts in which some of his friends and acquaintances got involved. Thus, he sought – largely in vain, as it turned out – to exert a calming influence on the famous humanist Francesco Robortello (1516–1567). Robortello, who was known as a "grammar dog" ("canis grammaticus") because of his fiery temper, engaged in a violent feud with Carlo Sigonio, whom Falloppia had known since his childhood days in Modena. The quarrel was, among other things, over the identification of the names and terms of office of the Roman consuls on then newly discovered but fragmented marble tablets. Sigonio had dared to accuse Robortello, quite rightly it appears, of various errors in this field and Robortello reacted deeply offended. He accused Sigonio, in turn, of numerous errors in an undertaking which did not present major difficulties for someone who was truly versed in Roman history, and which could only be excused by Sigonio's youth (Sigonio was in his thirties).[59]

Another scholarly controversy affected Falloppia much more directly and personally, namely that between Melchior Wieland and the famous botanist Pietro Andrea Mattioli. Mattioli was known for reacting to any criticism, whether it was justified or not, with vitriolic attacks *ad personam*. An "irascible naturalist" Henri Leclerc aptly called him.[60] In a detailed letter to the Swiss naturalist and botanist Conrad Gessner (1516–1565), Wieland had mentioned Mattioli in the context of his discussion of various plant names, such as "bulbocastaneus", "doronicum", and "moly", in a list of other "exceedingly learned" men and referred to him as the "god of herbalists" ("a Deo illo herbariorum Matthaeolo"). With regard to the identification of the "moly" of Dioscorides, however, he accused him of not having read the relevant passage in Galen, as he should have done, where the "moly" was clearly described.[61]

The letter appeared in print in 1557, along with a letter from Conrad Gessner, who in those years was engaged in another feud with Mattioli about the nature of *aconitum*. Gessner had found that the illustration of the *aconitum primum* in Mattioli's *Commentarii* was based only on the description Dioscorides gave of it and did not show the real plant, which, according to Gessner, looked quite different.[62] In a preface to the reader, the publisher Nicolaus Philesius explained that Gessner had shown him Wieland's letter, whereupon he asked Gessner for permission to print it together with a text by Gessner. According to Philesius, Gessner had initially refused because Wieland's letter was personal and private but had agreed in the end. Wieland himself later affirmed that his letter was indeed not intended for publication.[63]

In the next edition of his *Commentarii*, Mattioli retaliated with a vicious attack against both Gessner and Wieland. This "certain Prussian" ("Borussum"), whom he should better call a snake ("marassum"), had attacked the modest Mattioli, driven by "barbarian envy" with calumny and imposture and without any reason or understanding. For the time being, he could not even be bothered to reply and he left it open whether he would perhaps at some point "waste a few good hours" on refuting Wieland's claims.[64] In a personal letter to Aldrovandi, Mattioli followed up. Wieland had completely lost his mind. He had had to defend himself against the calumnies and the slander of this "barbaric, rabid, Prussian dog", whom he did not consider worthy of an answer.[65]

Mattioli never "honored" Wieland with a direct, public response. Instead, he published a long letter to Falloppia in which he complained once more about Wieland's "contumeliosam acerbitatem" and his "deliramenta". He compared him to a madman, who understood little, if not to say nothing at all, about medicinal plants. Scholars who returned to Prague from Padua, and knew of the friendship between Mattioli and Falloppia, had expressed their great astonishment that Falloppia shared the house with "this barbarous, godless, and almost inhuman Prussian". Just as they praised Falloppia, they detested Wieland. At times, Mattioli added, poisonous snakes were found among the most beautiful flowers and useless weeds sprouted in lush fields. He urged Falloppia to put Wieland in his place and dissuade him from his insults. If Wieland did not follow this advice, he would reveal his barbarism, his ignorance, and his inhumanity to the whole world.[66]

It is unlikely that Falloppia was impressed. Wieland's brief remark had been harsh but he had used a fairly moderate, civilized tone, in a letter, moreover, which, if Wieland and the publisher were to be believed, was not written for publication and was printed without his consent. It was Mattioli whose crude insults went far beyond the codes of civil conduct that governed in the contemporary republic of letters. Falloppia did not even reply to Mattioli's letter. And Mattioli certainly had underestimated Wieland. He had found his match. Wieland was an experienced botanist and very well versed in the botanical writings of the ancients. He immediately published a response.[67] The heading already announced a devastating broadside: "Refutation of Mattioli's calumnies and outline of about a hundred of his errors."[68] With his unbridled tongue, Wieland complained, Mattioli had seriously offended his name and now he, Wieland, would no longer mince his words either. Over almost forty pages, he relentlessly enumerated one error after another that he had found in Mattioli's new edition of the *Commentarii*. Even though Mattioli had repeatedly revised this unlearned and faulty "scrub",[69] it remained a terrible piece of work full of appalling errors. He also paid Mattioli back for the insult he had given him because of his origins: Mattioli had to learn that Nature had blessed the people north of the Alps with sharp minds, too. It was wrong to call them "barbarian", as Mattioli did. Rather, Mattioli himself was a most barbarian person, although he was born in the middle of Etruria.[70]

Mattioli was furious. This was no longer just a scholarly controversy about a couple of plant names. His reputation and authority had been massively attacked and this, according to respected contemporaries like Johannes Crato von Krafftheim, in some points quite rightly.[71] In a personal letter, Mattioli turned his wrath against Falloppia. For several months, he had waited in vain for a reply to his earlier letter, he complained. Falloppia had not even thanked him for the copy of the latest edition of his *Commentarii*. Instead, a friend from Padua had sent him the *Apologiae*, that diabolically infamous book by Falloppia's "pupil" Wieland. Wieland lived in Falloppia's house and ate his bread. He confided things to Falloppia that he told no one else. Undoubtedly, Falloppia therefore knew of Wieland's intentions and agreed with them. If Falloppia had really been that true friend of Mattioli's he had always pretended he was, it would have been his duty to restrain Wieland with all his authority and to prevent the publication. Instead, Falloppia had encouraged Mattioli to respond to Wieland's attacks. Apparently Falloppia wanted to damage Mattioli's honor and fame. Falloppia was the actual weapon and Wieland merely the projectile. Mattioli should have listened to the warnings of friends. But he had not imagined that Falloppia would betray him in such a way. Now he knew why Falloppia had sent him the letter of this "son of a priest and a public whore". No one should think that he would answer this rabid dog. To beasts like that one responded only with the sound of a good beating until their brain together with intellect fell into their mouth so they learned to talk more sense. This "seed of stench", who was worse than an animal, was wise to travel to Turkey. Otherwise, he would have been taught how to write and speak another language. Hopefully, God would inflict the punishment this scoundrel, this "sordid hermaphrodite", deserved for his misdeeds.[72]

In his correspondence with Aldrovandi, Mattioli followed up. He expressed his satisfaction that Aldrovandi was pleased with the letter to Falloppia which

Mattioli had published. Others, Mattioli claimed, had also sided with him. Only Falloppia had written to him in the meantime, claiming that everyone who had read this letter found it too violent, written with too much anger. But Falloppia had to be forgiven. Perhaps he loved the vices of "his Guilandino" and "the gallantry of such a kind hermaphrodite" more than the truth and Mattioli's virtue.[73] Mattioli added that some German students, on their return from Padua to Prague, had told him that "that beast" had claimed that he had found a letter in Falloppia's room in which Mattioli asked Falloppia to kill Wieland with poison. When challenged, Wieland could not show the letter, however, and claimed that the letter had been stolen from him. Indeed, according to Mattioli, Falloppia declared that no such letter would ever be found. The story is so implausible and Mattioli had given such good reasons for his answer with his insults to Wieland that Mattioli almost certainly made it up. It is true, however, that Mattioli had quite openly threatened Wieland with violence. He even still sought to harm him, when Wieland was in Turkey. He asked Aldrovandi repeatedly to get Antonio Cavalli to write to his father, Marino Cavalli. Marino Cavalli had been sent on a diplomatic mission to Constantinople and Wieland had traveled with him. Antonio was to tell his father about Wieland's shameful behavior, so that the father would be downright ashamed to have him in his own house. Aldrovandi should give a copy of Mattioli's letter to Falloppia to Antonio Cavalli to be forwarded to his father, so the father would know better about the origin, life, and customs of this cunning fellow.[74]

Aldrovandi had been in correspondence with Wieland for years.[75] His alleged approval of Mattioli's hateful public reply letter may well have been only a polite, diplomatic comment. At any rate, he apparently did not comply with Mattioli's request: in the following year, Wieland thanked Aldrovandi effusively for his letters of recommendation to the Venetian consuls instead.[76] Later Joachim Camerarius reported in a letter to Johannes Crato that Aldrovandi in fact was very indignant about Mattioli's letters and rather sided with Wieland.[77]

Mattioli still did not relent. In a personal letter to Crato that he wrote by hand in Crato's copy of his *Epistolae*, he did not attack Wieland directly but expressed his deep disappointment with Anguillara's new book. Anguillara had spent his life exploring plants but had nothing valuable to say. This was shown also by the fact that Anguillara never criticized Wieland but frequently wrote against Mattioli's opinions and covertly sought to offend him.[78] Mattioli also sought support from other quarters. In his *Epistolae* of 1561, he published a long letter to Johannes Crato, in which he replied to some of Wieland's points and accused him again of ignorance and slander.[79] In his reply, Crato criticized Wieland as excessively sharp and listed a number of points on which he disagreed with him. He also listed a number of other points, however, on which he himself had certain doubts regarding Mattioli's account.[80]

Wieland did not respond to Mattioli's published letter to Crato. His student Paulus Hessus, however, wrote a *Defensio* in which he commented on the twenty *problemata* that Mattioli, in Hessus' words, had picked out from the "very true" ("verissimis") findings of the "modest" Wieland. Before publishing his letter to Falloppia, Hessus declared, it would have been Mattioli's duty to ascertain whether Wieland had actually consented to the publication of his letter to Gessner. Instead, Mattioli had attempted in numerous letters to incite the hate of the entire European scholarly world against Wieland and had insulted him so foully that it would turn the stomach of any person with common sense. Wieland, Hessus rebuked Mattioli's insults *ad personam*, was of honorable origin and had a good education. He had not been banished and never worked as a donkey driver either.[81]

A "liaison dangereuse"?

In his attacks on Wieland, Mattioli used extremely foul language. As we have seen, he not only compared him to an animal but, in his letters to Falloppia and Aldrovandi, added allusions of sexual deviancy, calling him a "hermaphrodite", whose gallantry Falloppia perhaps loved more than the truth. In his letter to Aldrovandi, he also referred to him as "Trasoncolo", that is, little "Trasone".[82] Thraso, in Italian "Trasone", was one of the "flatterers" or "parasites" in Menander's drama "Kolax". Thraso also played a role in the comedy "Eunuchus" by Terence, which carries sexual references already in the title.

Unfortunately, we know little about Wieland's appearance. His portrait is found, in 1650, on the title page of the *Historia plantarum universalis* by Johann Bauhin and J. H. Cherler. It was copied several times but is unlikely to be authentic.[83] If it bears even a rough resemblance to the historical Wieland, he wore a full beard, which would not exactly suggest that he was a hermaphrodite. But Mattioli's slander was presumably not about an anatomical anomaly anyway. He clearly wanted to suggest a sexual relationship between Falloppia and Wieland, in which he saw Wieland playing the effeminate, passive part, which at that time was considered all the more despicable.

In 2012, Michele Visentin quoted Mattioli's letters in a carefully researched article for the journal *Il Mattino* in Padua. The website *Cultura-Gay.it* republished the piece, now with a bibliography where it can still be found today under the title *Falloppia & Guilandino. Una "liaison dangereuse" nella Padova del Cinquecento*.[84] Visentin refrained from speculating about a possible homosexual relationship between Falloppia and Wieland. In fact, no clear judgment is possible in retrospect. Research on the history of homosexuality has shown that our modern concept of a homosexual identity can be applied to earlier times to a limited extent only. Georg Handsch, for example, documented various sexual encounters with other men in his notebooks with a remarkable matter-of-factness,[85] but this does not mean that Handsch was "homosexual" in the modern sense. Moreover, sharing the household does not necessarily mean a high degree of intimacy. It was common, at the time, for students to live in the households of professors, and we hear repeatedly of visitors whom Falloppia hosted in his home. Some stayed for months or even years. Thus, Lodovico Carissimi, later lecturer on simples in Pavia, lived in Falloppia's house in 1554.[86] Around 1560, Falloppia's student Giovanni Battista Carcano Leone, who later taught anatomy in Milan, spent two years there.[87]

Apparently, the relationship between Wieland and Falloppia did raise some questions among their contemporaries in Padua, however. Mattioli, who lived in Prague at the time, and was well informed about events in Padua, repeatedly mentioned students from Prague who reported to him by letter or on their return from Padua about Falloppia and Wieland. It seems hard to imagine that Mattioli, who did not know him personally, would have given this specific direction to the savage insults he hurled at Wieland, if he had not heard rumors to that effect. And there is also some independent piece of evidence that the close relationship between Falloppia and Wieland raised some suspicion. In a letter he wrote to Joachim Camerarius after Falloppia's death, Georg Purkircher, then a student in Padua, reported that Wieland was unwilling to take on the lecture on the simples for the small salary offered and continued: "I hear that a new suitor of this Helena has been found." He promised to provide more details on another occasion but unfortunately no such letter kind is known to have survived.[88] By all appearances, some people in Padua at least perceived Falloppia as the previous "suitor", in line with Mattioli's insinuation that the "Helena" and "Trasoncolo" Wieland had the more feminine part in this relationship.

Final days

> May Gabriel Fallopus reach a Nestorian age,
> Who dissected the limbs of the human body more perfectly
> Than any other and assigned to the parts their respective uses.[89]

Falloppia's student Georg Handsch wrote this poem in the early 1550s. Presumably he hoped to win Falloppia's favor with it. A Nestorian age, however, was not to be granted to Falloppia. He died in 1562, not even forty years old.

Historians have largely agreed on the date of Falloppia's death, namely October 1562.[90] The date is confirmed by various contemporary sources. On 8 October, Leonora d'Este, who was in Falloppia's medical care in Padua at the time, wrote to her brother Alfonso that Falloppia was very sick. The next morning, on 9 October, at the sixteenth hour ("le 16 del IX"), she added a note to that letter before giving it to the messenger, who wanted to leave early, in which she told her brother that Falloppia "at this moment" ("in questo punto") was at the brink of dying and maybe had already died while the messenger was on his way to her.[91] In Padua as in Venice and in other areas of Italy, the hours of the day were counted from the time of the sunset. Accordingly, the sixteenth hour would have corresponded roughly to 9.30

a.m. today, with the sun setting around 5.30 p.m. in Padua on 9 October 1562.[92] The next day, on 10 October, Leon Crotto reported from Padua to Pietro Martir Cornacchia in the chancellery in Mantua that Falloppia had died after eight days of disease.[93] Favaro concluded from Leonora's added note that Falloppia died at the sixteenth hour of 9 October,[94] but clearly Leonora only expressed her fear that Falloppia might already have died in the meantime, based on what she had just learned about Falloppia's terminal state. A more precise indication comes from Barbieri's entry in the Modena council minutes. At the bottom of the minutes for 9 October, Barbieri added that Falloppia had died that day after the first hour of the night. Presumably, Barbieri was also using the "ora italiana", which would mean that Falloppia died at around 6.30 or 7 p.m. on 9 October.[95]

Generations of historians have speculated about the causes of his early death. Some authors, most notably Favaro, have described his death as the culmination of years of health problems, significantly promoted by the strains of his work as a practitioner and, in the cold winter season, as an anatomist.[96] Some authors have more concretely suspected pulmonary tuberculosis as the cause of death.[97] At that time, consumption was indeed a commonly diagnosed and widely feared cause of death. Considerable doubts remain, however. Falloppia's letters and other sources from the last years of his life do not mention the typical symptoms that were associated with consumption at the time and with pulmonary tuberculosis today, such as coughing up blood. In view of his father's colorful life, other authors have suspected that Falloppia suffered from congenital syphilis.[98] The early deterioration of his hearing could point in this direction but other typical consequences of a late form of syphilis have not been documented for Falloppia. And even if he did suffer from the disease: in Falloppia's days, the "French disease" commonly no longer took a fatal course.

In short, from today's point of view, the surviving sources offer no convincing evidence of any chronic, ultimately fatal ailment. It is true that Falloppia repeatedly mentioned illness episodes in his letters. In January 1554, Wieland reported to Aldrovandi that Falloppia had been ill in bed for sixteen days, with fever, violent pain in the stomach and the head, colics, vomiting, and other ailments; now he felt better.[99] In 1556, Falloppia complained that he had been in bed all winter until carnival.[100] In January of 1557, he cited a cold from which he had suffered as an excuse for not yet having answered the sick Corbinelli.[101] In the summer of the same year, again in justification of a belated reply, he complained of a fever and an epidemic catarrh that had affected many and confined him to bed for many days.[102] In none of these instances is there the slightest indication that Falloppia did not get better again. On the contrary, the symptoms and the limited duration of the various disease episodes suggest acute, transitory diseases. The only major piece of evidence that might point more concretely to a long-term deterioration of Falloppia's state of health is a letter to Aldrovandi in March 1557. He used to be full of fire, Falloppia complained, but now his labors ("fattica") had reduced him to a state in which he could hardly maintain himself. If he wanted to stay healthy, he could only eat once a day. His account served as a warning to Aldrovandi, however, who worked too hard and whom he admonished to take care of his own health; Falloppia also mentioned the robust Bartolomeo Maggi, who was killed by his labors.[103] He may therefore well have exaggerated. After all, in the spring of 1560, Falloppia still accompanied a Venetian embassy to Paris and only a couple of months before his death, he went to Ferrara and spent some time there. While consumption was known to often take a fatal course, he clearly saw no reason to fear his imminent death either. As we have seen, he announced a comprehensive anatomical work as late as 1561 and although he had neither wife nor children and thus no immediate heirs,[104] he did not even leave a will.[105]

The contemporary accounts of Falloppia's surprising, sudden death, as well as various other contemporary sources, also strongly suggest that he died of an acute illness. Barbieri noted in Modena that Falloppia had succumbed to "doglia di costa", that is, pain in the side (or ribs) and "febre pestilentiale".[106] According to Leon Crotto, he died within eight days "per la ponta".[107] Similarly, Sallustio Piccolomini, then ambassador of the Duke of Tuscany in Ferrara, gave the diagnosis of Falloppia's last illness as "mal di punta".[108] Presumably, this was the diagnosis at which the attending physicians in Padua had arrived, news of which had reached Ferrara and Modena. The terms "doglia di costa" and "mal di punta" or "ponta" were used largely synonymously with the more technical, scholarly term "pleurisy" ("pleuresia", or in Greek "pleuresis").[109] They all refer to a disease with a characteristic symptom, namely, a (usually unilateral) stabbing pain localized at the ribs, accompanied by fever. Today's physicians would most likely think of a local inflammation of the pleura, as it can occur in acute pneumonia.

On 12 October 1562, Falloppia was buried in the Basilica di Sant'Antonio in Padua. For some time after Falloppia's death, his tomb could be seen close to the entrance to the church.[110] It was later destroyed during building work and his bones were transferred to Melchior Wieland's tomb. The funeral speech was delivered by the then rector Jan Zamoyski (Zamoscius).[111] As Georg Purkircher (1530–1577) reported from Padua at the end of November 1562, unknown persons wrote a little poem: "Falloppia, you are not alone locked in this tomb. Together with thee lies equally buried our house." The author was possibly Melchior Wieland, who had lived in Falloppia's house for many years.[112]

The conflicts between Falloppia and his fellow professors did not end with his death. On the contrary, after Falloppia's death, according to the testimony of contemporaries, it became apparent just how great the hatred was. When he learned of Falloppia's death, Bassiano Landi, according to Purkircher, was said to have triumphed over it in a letter to Francesco Robortello.[113] Landi's joy was not to last long. Twelve days after Falloppia's death, he was attacked by two masked men. According to contemporary accounts, they used a spade and a firearm, inflicting three head wounds and shattering the bones of both arms.[114] Ten days later, on 31 October 1562, Landi succumbed to his injuries. During his funeral, someone posted a note on the gate of the Chiesa degli Eremitani, where Landi was buried, in broad daylight, with the words: "Alive, Landus, I was a plague, dying I was your death."[115] As Purkircher reported, it was generally believed in Padua that the word "plague" referred to Falloppia. Landi's letter to Robortello was believed to have fallen into wrong hands and people in Padua said that this put Landi in danger.[116] In a manuscript in the University Library of Erlangen, which gathers various lecture notes and other texts from the late sixteenth century, mainly from Padua, an unidentified scribe, probably retrospectively and from second hand, was even more outspoken. He claimed that Bassiano Landi "was miserably stabbed in Padua by his enemies, the disciples of Falloppia, because he seemed to rejoice in Falloppia's death".[117] Such violent revenge is not beyond the realm of the possible. Just a few weeks later, Purkircher reported about another murder in the environment of the university. Iacobus Cicuta, a count from Krk, was murdered by unknown persons in his house together with his servant.[118] The previous summer, his election as rector of the law students had failed due to violent protests because of his "impure life" and his "shameful infamies".[119]

Notes

1 Letter from Lelio Ruini to the *Quaranta* in Bologna, Padua, 7 June 1557, cit. in Costa, Ulisse Aldrovandi (1907), p. 82. The Ruini ranked among the leading families in Bologna.
2 Biblioteca Universitaria di Bologna, Mss Aldrovandi, 38², I, fol. 49 and fol. 50, letters from Falloppia to Aldrovandi, Padua, 10 November 1558 and 30 January 1559, ed. in Di Pietro, Epistolario (1970), pp. 43–44 and pp. 47–48.
3 Letter from Camillo Canonici to the *Quaranta* in Bologna, Padua, 11 February 1559, cit. in Costa, Ulisse Aldrovandi (1907), pp. 82–83.
4 Biblioteca Universitaria di Bologna, Mss Aldrovandi, 38², I, foll. 53–54, letter from Falloppia to Aldrovandi, Padua, 23 January 1561, ed. in Di Pietro, Epistolario (1970), pp. 55–56.
5 Biblioteca Universitaria di Bologna, Mss Aldrovandi, 38², I, fol. 57, letter from Falloppia to Aldrovandi, Padua, 10 October 1561, ed. in Di Pietro, Epistolario (1970), pp. 58–59.
6 Costa, Ulisse Aldrovandi (1907), p. 83, referring to a letter from Zambeccari to the *Quaranta*, 4 November 1561.
7 For Carcano's biography see Scarpa, Elogio (1813). He was born in 1536 in Milan and thus only about twenty-five years old at the time. He had studied medicine in Milan and had worked as a surgeon during a military campaign before he came to Padua and studied with Falloppia.
8 Carcano, Liber II (1574), fol. 36v; Scarpa, Elogio (1813), p. 12.
9 Panetto and Wiel Marin have suggested that Falloppia wanted to leave Padua also out of fear of the inquisition (Panetto and Wiel Marin, Falloppia (2001), p. 290); it seems highly unlikely, however, that Falloppia, if he had feared the inquisition, would have felt safer in Bologna which was under direct papal rule.
10 Biblioteca Universitaria di Bologna, Mss Aldrovandi, 38², I, foll. 53–54, letter from Falloppia to Aldrovandi, Padua, 23 January 1561, ed. in Di Pietro, Epistolario (1970), pp. 55–56.

11 The salary was increased to 270 fl. in 1560 but this was hardly enough to compensate for the massive increase of the cost of living in those years. On 21 June 1561, the Venetian authorities felt compelled to provide extra funds for the salaries of current and future professors in Padua, because, as they explained, prices had virtually doubled (Archivio di Stato di Padova, Archivio civico antico, Ducali 86, fol. 39r).

12 Biblioteca Marciana, Venice, Ms Ital. X, 142, foll. 62r–63r, *rotolo* (salary list) of the professors in the arts faculty for the academic year 1551/1552.

13 Tomasini, Gymnasium (1654), pp. 297–298.

14 Trincavella, Consilia (1587), col. 286 and col. 332.

15 Biblioteca comunale di Siena, Misc. XVI, C IX 32, foll. 1r–15v, notes by an unidentified student on three *collegia*.

16 Universitätsbibliothek Erlangen, Ms. 910, foll. 50r–53v.

17 Universitätsbibliothek Erlangen, Trew, Purkircher, Nr. 4, letter to Joachim Camerarius, dated 18 February 1563 (the date is probably wrong and should be March 18); for a detailed summary see www.aerztebriefe.de/00002573.

18 Falloppia, De vulneribus (1606), pp. 241–244.

19 Ibid., p. 241, "in colloquiis, et in alijs locis profiteantur se esse admodum exercitatos in re chirurgica, et recte hanc artem callere, tamen oppositum postea ostendunt, dum asserunt excrementum copiosum in vulnere gigni ob laesam facultatem".

20 ÖNBW, Cod. 11224, fol. 145v: "Quia nemo homo peritus anatomes est, qui non percipiat nervos opticos habere foramina, reliquos vero esse solidos." The quote is from Handsch's notes on some lectures by Bassiano on the eye. They follow his notes on Bassiano's lectures on Galen's *Ars parva* in 1552/53 but Handsch mentions that they were from a previous year.

21 ÖNBW Cod. 11210, fol. 24v.

22 Tomasini mentions manuscript notes on Landi's anatomical lectures in 1549, which by Tomasini's time were in the possession of Ettore Trivisano (Tomasini, Bibliothecae (1639), p. 114, on "Anatomica quaedam e lectionibus Bassiani Landi. Collecta Anno 1549 a Bernadino Trivisano"). Georg Hieronymus Welsch, in the seventeenth century, owned a manuscript with notes on sixteen anatomical lectures by Landi but it is unclear whether Landi held them before or after Falloppia starting teaching (Welsch, Consiliorum (1676), note on p. 236).

23 Landi, *De humana historia* (1542); see also Ongaro, Bassiano Landi e Andrea Vesalio (1998), p. 38.

24 Biblioteca Universitaria di Bologna , Mss Aldrovandi, 38[2], I, fol. 49, letter from Falloppia to Aldrovandi, Padua, 10 November 1558, ed. in Di Pietro, Epistolario (1970), pp. 43–44.

25 It is possible that Fracanzano was referring to the observation of cases of what is known today as vaginal atresia.

26 ÖNBW, Cod. 11210, fol. 158v.

27 SUBG, Ms. Meibom 20, foll. 193v–194r.

28 The Helmstedt Anonymous used the expression "propter tumultum et petulantiam scholasticorum". The students continued to remain very noisy when Oddo was speaking, which suggests that the noise was not an expression of protest against Falloppia.

29 Ibid., fol. 194r.

30 Ibid., "propter consuetos clamores scholasticorum non est responsum".

31 Ibid., fol. 188r–v.

32 Ferinarius, Narratio (1601), no pagination.

33 Letter from Bonifaz Zwinger to Theodor Zwinger, Padua, 26 February 1587, Universitätsbibliothek Basel, Frey-Gryn Mscr. I 11, fol. 417r (digital image: https://www.e-manuscripta.ch/bau/content/pageview/177268).

34 Eriksson, Vesalius' first public anatomy (1959).

35 ÖNBW, Cod. 11210, fol. 187r and fol. 191v; see also Mache, Anatomischer Unterricht (2019). Ap(p)ellato was professor of *medicina practica* in Padua since 1543; in November 1558 Falloppia reported his death.

36 Decree of the Venetian Senate, 23 September 1551, ed. in Tomasini, Gymnasium (1654), p. 96.

37 ASV, Senato I. R o 38. Terra. September 1551-September 1552, fol. 163v (and copy of the same document in Archivio antico dell'Università di Padova, n. 665, fol. 36r), ed. in Favaro, Gabrielle Falloppia (1928), pp. 216–217: "obbligo di legger la chirurgia, simplici, et di tagliar, legger, et mostrar l'anatomia, sicome ha fatto l'anno passato."

38 Georg Handsch's notes on Landi's lectures on the eyes and the physiology of vision may reflect some of this anatomical teaching (ÖNBW, Cod. 11224, foll. 138r–149v).

39 Letter from the *Riformatori*, 30 December 1553, "per haver letta certa lettion de annatomia"; ed. in Rippa Bonati, Su un insegnamento (1998), pp. 55–61.

40 Archivio antico dell'Università di Padova, n. 675, fol. 165r, 6 December 1554; ed. in Favaro, Gabrielle Falloppia (1928), p. 223.

41 Archivio antico dell'Università di Padova, n. 675, foll. 171r–v, Padua, 25 January 1555 ("Di Padova il XXV di Gennaro 1555" but "Ex offitio nostro artistarum die XXI Januarij 1555"), ed. in Favaro, Gabrielle Falloppia (1928), pp. 224–225.

42 Ibid.

43 Schieß, Briefe (1906) p. 20.

44 ASV, Riformatori 63, letter from Falloppia to the *Riformatori*, Padua, 12 December 1556.

45 Letter from Falloppia to Lodovico Corbinelli, 29 January 1557, ed. in Angelini, Una lettera (1900), reprinted, without access to the original, in Di Pietro, Epistolario (1970), pp. 31–32.

46 Letter from the *Riformatori* to the rectors, Venice, December 7 and December 15, 1557 in ASV, Lettere dei Riformatori allo Studio, 1555–1559, filza 63; ed. in Favaro, Gabrielle Falloppia (1928), p. 228.

47 On Wieland (the Italian "Guilandino" derived from the latinized "Guilandinus") see De Toni, Melchiorre Guilandino (1921); Herrmann, Ein Preuße (2015).

48 On Cavalli see Olivieri, Cavalli (1979); for much of his life, he served as a Venetian ambassador but in the 1550s, he was one of the Venetian *Riformatori* in charge of Padua university and later both he and Wieland were in Constantinople.

49 According to Visentin, Una "liaison dangereuse" [2012] and Herrmann, Ein Preuße (2015), Falloppia lived in a house in the *Contrada delle Beccherie*, today's *Via Cesare Battisti*. Both authors do not provide a reference and so far I have not found independent evidence for this claim. According to Favaro, Gabrielle Falloppia (1928), p. 122, Falloppia does not figure in contemporary documents on house ownership in the Padua town archives. Falloppia's property at the time of his death included a house in Modena but none in Padua, which suggests that the house in Padua was rented.

50 Wieland, Apologiae (1558), fol. 17v; the letter is dated "IX. Kal. Mar. MDLIIX" (21 February 1558).

51 Letter from Mattioli to Aldrovandi, 26 November 1558, ed. in Raimondi, Lettere (1906), pp. 158–163.

52 Letter from Wieland to Aldrovandi, 9 June 1559, Biblioteca Universitaria di Bologna, Mss Aldrovandi, 38^2, I, foll. 135-136; ed. in Fantuzzi, Memorie (1774), pp. 219–221.

53 De Toni, Melchiorre Guilandino (1921).

54 Favaro, Gabrielle Falloppia (1928), p. 131; Favaro does not explain why Falloppia would have had to travel to Greece to free Wieland from the hands of pirates who operated from northern Africa.

55 Wieland, Papyrus (1572), p. 110.
56 De Toni, Melchiorre Guilandino (1921).
57 Letter from Falloppia to Ulisse Aldrovandi, Padua, 20 September 1554, Biblioteca Universitaria di Bologna, Mss Aldrovandi, 38^2, I, fol. 44, ed. in Di Pietro, Epistolario (1970), pp. 25–26.
58 Letters to Alfonso II, 31 August 1560 and 15 November 1560, ed. in Di Pietro, Epistolario (1970), pp. 51–52 and p. 53 (the original manuscripts are in the Archivio di Stato di Modena, Arch. per materie: Medici e medicina, Documenti esposti all'Esposizione Nazionale di Torino (1898)); Foucard, Documenti (1885), p. 23, offers a photographic reproduction of the second letter. It is not entirely certain that Falloppia was referring to Anguillara. He only wrote of "Messer Aloise" – not "Luigi", as he had called Anguillara in 1554 in a letter to Aldrovandi. "Luigi", "Aluigi", "Alvise", and "Aloysius" were variants of the same name which was very common. One of Falloppia's collegues in Padua was Alvise Bellocati. If Falloppia actually recommended another "Aloise", he would have sought to prevent Anguillara from being employed in Ferrara and thus in the service of his own sovereign. However, this is highly improbable, since Anguillara was, in fact, appointed.
59 Robortello, De convenientia (1557), introduction, no pagination; Sigonio, Emendationum (1557).
60 Leclerc, Un naturaliste irascible (1927).
61 Wieland, Letter (1557), p. 22.
62 Delisle, Letter (2004).
63 Wieland, Apologiae (1558), fol. 3v.
64 Mattioli, Commentarii (1558), pp. 541–542, text appended to the chapter on *aconitum.*
65 Biblioteca Universitaria di Bologna, 38^2, I, fol. 19, letter from Mattioli to Aldrovandi, Prague, 29 January 1558, ed. in Raimondi, Lettere (1906), pp. 155–158.
66 Mattioli, Epistola (1558); also in Mattioli, Epistolarum (1561), pp. 159–172, "hunc barbarum, impium, ac pene immanem Borussum"; "suam barbariem, imperitiam, inhumanitatemque".
67 Wieland, Apologiae (1558).
68 Ibid., fol. 3r: "Matthaeoli calumniae refutatae, atque eiusdem centum circiter errores obiter indicati".
69 Ibid., foll. 4r and 16r–v.
70 Ibid., fol. 18r; Mattioli was from Siena, which was in the area of ancient Etruria.
71 Mattioli, Epistolarum (1561), pp. 271–273, with Mattioli's response (ibid., pp. 273–275).
72 Autograph copy of a letter from Mattioli to Falloppia, Biblioteca Universitaria di Bologna, Mss Aldrovandi 38^2, I, foll. 20–21, ed. in Di Pietro, Lettere (1970), pp. 44–47; in a letter to Aldrovandi (26 November 1558, ed. in Raimondi, Lettere (1906), pp. 158–163), Mattioli mentioned enclosing a copy of his letter to Falloppia, which would explain why it ended up among Aldrovandi's correspondence. The copy does not indicate the date and Di Pietro assumes that the letter was written in late 1558. The *Apologia* which had apparently only just been sent to Mattioli when he wrote this letter would suggest an earlier date, however, quite possibly already in late spring. By November 26, when Mattioli sent the copy of his letter to Falloppia, Wieland had already left Padua, as Mattioli himself indicates.
73 Letter from Mattioli to Aldrovandi, 26 November 1558, Biblioteca Universitaria di Bologna, Mss. 38^2, I, foll. 22–23; ed. in Raimondi, Lettere (1906), pp. 158–163.
74 Ibid; Mattioli returned to this request in June 1559 and wanted to know whether Aldrovandi had made Antonio write another letter to his father and whether the

father had responded (ibid., pp. 163–164). By the fall of 1559, Mattioli claimed that Wieland, who was on his way to Jerusalem, was doing everyhing through friends to win his friendship – a rather implausible assertion, to say the least (Biblioteca comunale Aurelio Saffi, Forlì, Autografi Piancastelli 1431, letter from Mattioli to Ulisse Aldrovandi, 20 September 1559, ed. in Raimondi, Lettere (1906), pp. 164–165).

75 De Toni, Spigolature XI (1911).
76 Letter from Wieland to Aldrovandi, Cairo, 9 June 1559, Biblioteca Universitaria di Bologna, Mss Aldrovandi, 38^2, I, foll. 135–136; ed. in Fantuzzi, Memorie (1774), pp. 219–221.
77 Biblioteka Uniwersytecka, Wrocław, R 246, Nr. 104 (new: 102), foll. 137–138, letter from Joachim Camerarius to Johannes Crato von Krafftheim, Bologna, 28 September 1561 (detailed summary by Ulrich Schlegelmilch under www.aerzte briefe.de/id/00009901): "epistolis Matthioli admodum est offensus et videtur inclinare ad partes Guilandini".
78 Library of the Národní Muzeum, Prague, Nostická knihovna, H 365, letter (in Italian) from Pietro Andrea Mattioli to Johannes Crato von Krafftheim, Prague, 8 February 1561.
79 Mattioli, Epistolarum (1561), pp. 239–234, not dated; reprinted by Hessus as Mattioli, Adversus XX. problemata (1562).
80 Mattioli, Epistolarum (1561), pp. 271–273, with Mattioli's response (ibid., pp. 273–275).
81 Hessus, Defensio (1562).
82 Letter from Mattioli to Aldrovandi, 26 November 1558, Biblioteca Universitaria di Bologna, Mss. 38^2, I, foll. 22–23; ed. in Raimondi, Lettere (1906), pp. 158–163.
83 De Toni, Melchiorre Guilandino (1921), p. 76, mentions a water color portrait in the botanical institutes in Bologna and Padua, which I have not yet been able to see.
84 Visentin, Una "liaison dangereuse" [2012].
85 E.g. ÖNBW, Cod. 11183, fol. 59v: "Ter manuduxi cum Venceslao Sseliha, in Maio"; "manuductio" was Handsch's term for masturbation.
86 Di Pietro, Epistolario (1970), note on p. 33.
87 Carcano, Liber II (1574), fol. 36v.
88 Universitätsbibliothek Erlangen, Trew, Purkircher, Nr. 3, letter from Georg Purkircher to Joachim Camerarius, Padua, 27 January 1563: "Lectionem simplicium fastidit Guilandinus pro hoc stipendio. Audio huius Helenae novum procum inventum de quo alias."
89 ÖNBW, Cod. 11210, fol. 186v: "Viuat Nestoreos Gabriel Fallopus [sic] in annos/ Corporis humani quo nemo rectius artus/ Dissecuit, certos assignans partibus vsus."
90 Favaro, Gabrielle Falloppia (1928), pp. 113–115 and p. 143, with a footnote that lists various older writings most of which agree on this date; Di Pietro, Contributo (1974), p. 64.
91 Ed. in Campori/Solerti, Lucrezia (1888), p. 145.
92 Sunset time for Venice, calculated by www.world-timedate.com.
93 Archivio di Stato di Mantova, Archivio Gonzaga, E, XLV, 3, busta 1495, ed. in Favaro, Gabrielle Falloppia (1928), p. 234.
94 Favaro, Gabrielle Falloppia (1928), p. 143.
95 Di Pietro, Contributo (1974), p. 63.
96 In his biography of Falloppia, Favaro (Favaro, Gabrielle Falloppia (1928)), frequently underscores Falloppia's declining health. But his account of Falloppia's medical condition needs to be read with caution. In one of the most doubtful passages of his study, he even speculates – without any plausible evidence – that Falloppia may have suffered from genital hypoplasia (ibid., p. 128).

97 O'Malley, Falloppia's account of the orbital muscles p. 77.
98 Favaro, Gabrielle Falloppia (1928), p. 154.
99 Letter from Wieland to Aldrovandi, 4 January 1555, ed. in De Toni, Spigolature (1911), pp. 158–160.
100 Biblioteca Universitaria di Bologna, Mss Aldrovandi, 38^2, I, fol. 43, letter from Falloppia to Ulisse Aldrovandi, Padua, 29 May 1556, ed. in Di Pietro, Epistolario (1970), pp. 27–28, "che non habbia bisogno di donne intorno, che lo governino, perchè io non ho donne in casa a ciò".
101 Letter from Falloppia to Lodovico Corbinelli, Padua 29 January 1557, ed. in Angelini, Una lettera (1900), reprinted, without access to the original, in Di Pietro, Epistolario (1970), pp. 31–32.
102 Biblioteca Universitaria di Bologna, Mss Aldrovandi, 38^2, I, fol. 47, letter from Falloppia to Aldrovandi, Padua, 16 March 1557, ed. in Di Pietro, Epistolario (1970), pp. 35–36.
103 Biblioteca Universitaria di Bologna, Mss Aldrovandi, 38^2, I, fol. 45, letter from Falloppia to Aldrovandi, Padua, 16 March 1557, ed. in Di Pietro, Epistolario (1970), pp. 33–34:

> Specchiatevi nel povero Maggio il quale più robusto di V. S. fu dalla fattica ucciso, specchiatevi in me il quale era tutto fuoco et dalla fattica sono ridotto a mal termine in guisa che se voglio star sano mi bisogna mangiare una volta sola al giorno, et non essere huomo quasi, et con stento anchor mi mantengo.

104 Poem by Casimirus Accursius, in: Zamoyski, Oratio (1562), fol. 9v: "Fallopi proles tibi non est ulla superstes."
105 Di Pietro, Contributo (1974), pp. 63–64. His fortune, which was estimated to amount to 8,000 to 9,000 scudi, fell to Francesco Falloppia.
106 Cit. in Di Pietro, Contributo (1974), p. 63.
107 Archivio di Stato di Mantova, Archivio Gonzaga, E, XLV, 3, busta 1495, ed. in Favaro, Gabrielle Falloppia (1928), p. 234.
108 Campori, Lettere (1864), p. 434, referring to a letter to the Duke, Ferrara, 22 October 1562; Campori found the letter in the "central archive" in Florence but did not provide any more detailed information.
109 Kramer, Dictionarium (1702), p. 919.
110 Falloppia, Opera (1585), vol. 2, printer's letter to the reader.
111 Zamoyski, Oratio (1562).
112 Universitätsbibliothek Erlangen, Trew, Purkircher, Nr. 2, letter from Georg Purkircher to Joachim Camerarius, Padua, 26 November 1562: "Falloppi hoc tumulo solus non clauderis, una/ Est pariter tecum nostra sepulta domus." In 1563, Giovanni Rossettino bemoaned Falloppia's death in a long poem (Rossettino, Canzone (1563)).
113 Universitätsbibliothek Erlangen, Trew, Purkircher, Nr. 2, letter from Georg Purkircher to Joachim Camerarius, Padua, 26 November 1562.
114 Biblioteca Ambrosiana, Milan, G. 273 inf., lettera 113, fol. 215r, letter from Vincenzo Guarino to Alvise Mocinego, 22 October 1562, cit. in Ferretto, Maestri (2012), p. 20; Ferretto does not link the murder with Falloppia's death but suggests a connection with Landi's position in the debate on the immortality of the soul.
115 Universitätsbibliothek Erlangen, Trew, Purkircher, Nr. 2, letter from Georg Purkircher to Joachim Camerarius, Padua, 26 November 1562: "Pestis eram vivus, moriens fui mors tua Lande. Pestem istam omnes Fallopium exponunt"
116 Universitätsbibliothek Erlangen, Trew, Purkircher, Nr. 3, letter from Georg Purkircher to Joachim Camerarius, Padua, 27 January 1563:

> Pestis eram vivus, moriens fui mors tua Lande. Id omnes de Fallopio interpretantur, de cuius morte quia Bassianus, ut dicitur, per literas apud Robortellum

tunc Venetiis quaedam sua negotia curantem triumphavit, ideo cum literae interceptae essent in hoc periculum incidit.

117 Universitätsbibliothek Erlangen, Ms. 908, fol. 181v, possibly a copy of an earlier text: "Miserrime confossus est Pataviis a suis adversariis Fallopii discipulis [...], propterea, quod laetari videbatur in funere Fallopii." The text wrongly gives the year of Landi' death as 1561 and describes the anonymous text which was posted during Landi's funeral as a threatening announcement already made during that of Falloppia: "Unde quidam in funere Fallopii hunc versum publice affixisse fertur: Mors [sic] eram vivens, moriens ero mors tua Lande." The inaccuracies suggest that the account was based on later hearsay only.
118 Universitätsbibliothek Erlangen, Trew, Purkircher, Nr. 5, letter from Georg Purkircher to Joachim Camerarius, Padua, 20 February 1563; cf. www.aerzte briefe.de/id/00003018 (summary by M. Clement).
119 Tomasini, Gymnasium (1654), p. 412.

7 Legacy

Students and successors

Falloppia's career as a university teacher lasted only for about a decade and a half but especially in Padua he must have taught and left his mark on literally hundreds of prospective physicians from all over Europe. Among them, we find people like Georg Handsch who later served as a personal physician to Archduke Ferdinand[1] as well as personalities like Volcher Coiter, who played a major role in boosting the status of surgery in the German lands,[2] Giovanni Battista Carcano Leone (1536–1606), later professor of anatomy in Pavia,[3] and Girolamo Fabrizi d'Acquapendente, who together with his student and – soon – rival Giulio Casseri continued to assure to Padua a leading role in anatomy and surgery.[4]

It took some time to fill the gap left by Falloppia's death. His combined professorship of anatomy, surgery, and simples was split up. In December 1562, Francesco Lendinara was appointed to perform the dissections for a salary of 35 fiorini, while Alvise Bellocati was entrusted with the actual anatomical lecturing. Wieland, who had been made prefect of the Botanical Garden in the meantime, was to take over the lecture on the simples but hesitated at first to accept the (additional) position for the small salary he was offered.[5] In April 1565, Girolamo Fabrizi d'Acquapendente was finally appointed to succeed Falloppia as a lecturer of anatomy and surgery. He had studied medicine in Padua in the 1550s, possessed, like Falloppia, great manual dexterity, and was at the same time an experienced and sought-after surgeon.[6] Much more than Vesalius and even than Falloppia, he saw the study of anatomical structures above all as a prerequisite for the understanding of physiology, of the functions of the body, or, in his own terms, of the faculties of the soul. While the physiology of Vesalius and Falloppia was framed above all by Galenic teleology,[7] Fabrizi made the Aristotelian doctrine of the soul his central point of reference.[8] He extended his anatomical research to animals to study and show how Nature provided animals with very different tools to achieve the same purposes in the service of the faculties of the soul. Among the large-format anatomical color plates that Fabrizi (or his student and then competitor Giulio Casseri)[9] commissioned, there

DOI: 10.4324/9781003242000-8

are superb pictures of high aesthetic quality depicting the anatomy of various animals.[10]

Fabrizi's teaching with its focus on individual parts of the body and their role as tools of the various faculties was controversial. There were complaints and protests.[11] The students wanted and expected to witness a systematic anatomical demonstration of the whole human body, which would provide them with the knowledge they needed as medical practitioners. Konrad Zinn, who studied in Padua in the 1590s, reported that students at that time already preferred the teaching of Giulio Casseri. Zinn himself attended only one complete dissection during his time in Padua, and it was performed by Casseri, not by Fabrizi.[12]

Posthumous publications

Although Falloppia had illustrious successors, interest in his teachings and findings remained very much alive. Soon after his death, some of his students started publishing their notes on his lectures on various topics. Petrus Angelus Agathus and Andrea Marcolini were particularly active here. Clearly, former students and publishers perceived a market for these notes and they continued their efforts for decades after Falloppia's death. From the 1580s, the publication of student notes on Falloppia's lectures on individual subjects, or in the case of Agathus' edition of the *Opuscula* on several topics, culminated in editions of his collected "works".

The willingness of the printers to take the risk of such costly editions suggests a persistent demand. They also got into fights which, at the same time, shed light on the strategies and practices used by printers in the competitive and apparently lucrative market for medical publications in general. In 1584, some twenty years after Falloppia's death, a complete edition of Falloppia's *Opera* was published by Valgrisi in Venice. In addition to Falloppia's *Observationes anatomicae* and his letter to Girolamo Mercuriale *De asparagis*, it offered student notes on *De simplicibus medicamentis purgantibus, De compositione medicamentorum, De medicatis aquis, De metallis et fossilibus, De ossibus, De cauteriis, De vulneribus capitis, De vlceribus, De tumoribus praeter naturam, De morbo gallico*, and, also based on student notes, the *Institutiones anatomicae*, and Michinus' five *Observationes anathomicae* now – maybe to avoid confusion with Falloppia's own treatise – under the title *Obseruationes de venis*.[13] In the same year, Andreas Wechel's heirs in Frankfurt came out with their own largely identical edition, which by all appearances plagiarized the Venetian edition. The publishers even reprinted Valgrisi's dedicatory epistle to the Paduan patrician Giacomo Antonio Cortuso.[14]

In 1585, a second Frankfurt edition offered the same texts but was now complemented by a second volume, edited by Giovanni Pietro Maffei, a surgeon from Treviso, which comprised *De medicamentis simplicibus, De materia medicinali* (commentary on the first book of Dioscorides' *Materia medica*), *De luxationbus, De fracturis, De partibus similaribus, De vulneribus in genere*, *De vulneribus particularibus* (on injuries of the eyes and other parts of the body), *De decoratione*, and some additional chapters on preternatural tumors, which were missing in the 1584 edition.[15] This 1585 edition carried a six-year general imperial printing privilege for Falloppia's work from November 1582.

In the place of the dedicatory epistle by Valgrisi, the publishers now printed a letter from Johannes Crato, a personal physician to the Emperor, to his colleague Jacobus Scutellarius.[16] This letter has led some historians and librarians to the false conclusion that Crato was the editor. As Crato explained to Scutellarius, he had written a letter of recommendation for Falloppia's collected works in the previous year, at the request of the publisher Aubrius, in which he praised Falloppia's work and warned of the dangers of chemical medicines. However, he had done this hastily and the letter fortunately was not published. The letter to Scutellarius in the 1585 edition was a letter about Falloppia's work, not a dedicatory epistle, although the publishers printed it with the running title "epistola dedicatoria". As Crato explained, he hesitated to say anything about a deceased man that could be taken as criticism but one should also not remain silent when someone lead others into error. He had never heard Falloppia teach and he had only had access to Falloppia's work through a fellow student. Two things in particular appealed to him, apart from Falloppia's anatomical skill and his discoveries, namely his effort to find good, not commonly used medicines and his study of metals.

In 1600, the Frankfurt publishers reprinted their edition, again with the letter by Crato and even with the imperial privilege of 1582.[17]

In 1606, Giovanni Antonio and Giacomo de Franciscis in Venice published their own three-volume edition of Falloppia's works, with the 1585 letter from Crato to Scutellarius but claiming that their edition was very different from the Frankfurt edition.[18] According to the publishers' letter to the reader, it was based on the notes of Andrea Marcolino, one of Falloppia's favorite students (some of whose notes had already found their place in the previous editions from Frankfurt and Venice). Marcolino, they explained, had given their late father seventeen years ago two manuscript volumes with his notes not only on Falloppia's public lectures in Padua but also of other things he had from Falloppia and that probably no other student had ever seen. Falloppia had been eager to share his secrets with Marcolino, who already prepared his notes for printing and presented them to Falloppia while he was still alive. Falloppia had praised them, they claimed, and had welcomed a publication but asked that they not be printed until after his death. After Falloppia's sudden death, Marcolino had not pursued this matter and the father of the two brothers, when he finally received the manuscripts, did not get around to printing the texts either, they claimed. In the meantime, the Frankfurt edition had come out, however, and here the two brothers brought out the big guns. Their edition, they declared, was very different and far superior from the one that had been published in Frankfurt. The hands of strangers had falsified what Falloppia dictated, omitting or neglecting important parts and adding other useless, vain, false things, so that Falloppia would hardly recognize his work if he returned to the living. Not all tracts of the Frankfurt edition were to be rejected but those on external, surgical diseases were blatantly imperfect, disorganized, garbled, confused, and mutilated they claimed.[19]

The Frankfurt publishers were fuming. In 1606, they brought out an *Appendix* to their 1600 edition.[20] In the preface, they complained bitterly about the cunning, greed, and malice of the Venetian publishers, who did not respect agreements. The Venetians had learned of their intention to publish a new edition in Frankfurt and had offered them further, previously unpublished lecture transcripts for sale and they had bought them. The Venetians had sworn that this was all that was known of Falloppia's surviving works. Presumably, the Frankfurt publishers were referring to the 1585 edition: as outlined above, the first volume of the Frankfurt edition included the same texts as those of 1584, while the second volume offered various new texts, mostly on surgical topics; the 1600 edition was only a reprint. The Frankfurt publishers declared the slander by the Venetians to be a perfidious lie. Their edition differed from the recent Venetian one only in the arrangement of the titles – and, as they admitted – "a few additions", texts, it is implied, whose existence had been concealed from them. The *Appendix* now offered those missing texts: above all additional chapters on Falloppia's lectures on various types of ulcers and on certain kinds of dislocations, a few pages on head wounds and Falloppia's *De modo consultandi*, altogether about ninety pages. Thus completed, the publishers declared their edition should be more welcome to readers than the new Venetian one, which, they returned the slander, was so bad that one would find almost more errors than syllables.

The book of secrets

The posthumous publications of Falloppia's lectures seem to have enjoyed a considerable and lasting success and popularity. Numerous copies even of the large and costly folio editions of his *Opera* have come down to us, in libraries all over Europe, some with an expensive, elaborate leather binding. The most successful and most widely circulating work, however, that appeared under Falloppia's name, was not devoted to anatomy or surgery. It was a collection of "secrets", of recipes for remedies that were said to possess particular curative powers in certain diseases, and of instructions for the preparation of other substances that could be useful in the household, for example. The book first appeared in 1563, in Italian, under the title *Secreti diversi et miracolosi*.[21] Further Italian editions are documented for 1565, 1570, 1578, 1580, 1582, 1585, 1597, 1611, 1620, 1640, 1650, 1664, and 1676[22] and the list may not even be complete. In 1571, 1573, and perhaps also again in 1580, a German edition was published in Augsburg under the title *Kunstbuch*. Further German editions followed in Augsburg in 1584, 1593, and 1597, and in

Frankfurt in 1616, 1641, and 1651.[23] Moreover, probably around 1580, Friderich Gutknecht brought a selection of fifty selected recipes of the "most eminent master" Falloppia into print.[24] In 1690 and 1715,[25] significantly expanded versions of the *Kunstbuch* appeared under titles such as *Gabrielis Falloppi neu eröffnete vortreffliche und rare Geheimnisse der Natur*. Extant copies show that the *Secreti* also found its way into the monastery libraries.[26]

Publishers underscored Falloppia's authorship. "Published by the author himself in Italian" the frontispiece of the *Wunderbarlicher und menschlichem Leben gewisser und sehr neu nutzlicher Secreten drey Bücher* (Frankfurt 1616) read. The Frankfurt edition of 1715 even prefixed an alleged author's portrait (see Figure 7.3), which, according to the caption, depicted Falloppia (who died at the age of 39) as a physician and astrologer at the age of 73.[27] Contemporaries already expressed their doubts about Falloppia's authorship, however. Andrea Marcolino declared soon after the *Secreti* first came out that the attribution to Falloppia was completely wrong ("falsissime").[28] Historians have unanimously agreed with this judgment, and for good reasons. In Falloppia's days, physicians usually wrote their prescriptions in Latin and not in the vernacular, as in the *Secreti*. Moreover, the book by no means only offered medical prescriptions in the narrower sense. The readers could also find, for example, remedies for the treatment of animals, instructions for the production of ink, and the advice to throw a piece of bread into the water in order to locate a drowned person (it would immediately attach itself to his or her body). This was typical of the genre of *secreta*, which was very popular at the time,[29] while it is virtually out of the question that Falloppia passed on such non-medical recipes to his students. Even the publisher of the first edition of the *Secreti* of 1563, Marco di Maria, was remarkably cautious: Falloppia, he claimed, had received these recipes from various writers. They were, he implied, not Falloppia's own.[30]

In historical research, Petrus Angelus Agathus has widely been considered the true author or compiler of these *Secreti*.[31] I have not found any evidence of Agathus' involvement, however, and the attribution is probably due to a misunderstanding. In the *Opuscula*, his edition of various smaller works by Falloppia, Agathus included an *Arcanorum liber* of about fifteen pages. Here, he assembled a few dozen "secreta et experta medicamenta" and referred, in the introduction, to further recipes that could be found in *De morbo gallico* and other works by Falloppia.[32] Agathus by no means published all these *arcana* as Falloppia's, however. He explicitly identified about a dozen recipes that he owed to him and distinguished them from those he had from other authors, such as Elideo Padovani,[33] Francesco Frigimelica, and Girolamo Capivaccio, who enjoyed a good reputation as practitioners in Bologna and Padua respectively, at that time, and he attributed other recipes explicitly to "uncertain" authors.

The recipes Agathus explicitly attributed to him may very well have come from Falloppia's mouth. Most of them were for the treatment of surgical ailments, of ulcers, fistulas, and wounds, including those from the bite of a rabid dog, of scabies, leprosy, leprosy, and burns. In addition, there were remedies against gonorrhea, impotence, and "tooth worms", and a sleeping pill. Some recipes offer fairly detailed instructions on practical procedures for the treatment of various surgical ailments rather than recipes in the proper sense. For example, the instructions for gluing the edges of wounds together in patients who did not admit suturing, one reads: "I use very simple glue. I take egg white and beat it very carefully and put it in a large tin bowl, on top of it a tissue and then slaked lime, as a very fine flour." This was very much in the style we know from student notes on Falloppia's lectures. The recipe went on to describe how one then had to carefully blow away excess powder and spread the medicine around the wound and finally suture it.[34] To all appearances, Agathus reproduced here recipes and remarks the students heard from Falloppia in lectures or when they saw patients. Especially since he only attributed some of the recipes to Falloppia but others explicitly not, these recipes were very likely indeed authentic.

I have not found these – very few – recipes, which Agathus explicitly attributed to Falloppia in the various editions and translations of the *Secreti*. Even where the *Secreti* describe remedies such as terebinth oil or sleeping pills, which Falloppia also mentioned in his lectures, the recipes are very different. The true author or compiler of the *Secreti* thus remains in the dark.

Rather than offering an insight into Falloppia's actual medical practice, the *Secreti* and their enduring popularity are above all revealing of the image that contemporaries and subsequent generations had of him. Clearly, publishers felt that Falloppia's great renown as a practitioner would help push the sales. And thanks to the outstanding success of the *Secreti*, in turn, Falloppia became and continued to be a familiar name among medical laypersons, not as an anatomist but as a source of superior secret knowledge, especially about the preparation of effective medicines.

Portraits

Visual evidence for Falloppia's lasting fame comes from the various portraits, which claim to represent Falloppia, which look quite different and are probably all posthumous and not true to life but fiction.[35] At least one portrait must have been painted or drawn during his lifetime or shortly after his death. Melchior Wieland, who lived in Falloppia's house for years, mentioned a portrait of his friend that he had in his study ten years after Falloppia's death.[36] Wieland did not describe it. We therefore do not know whether this – presumably realistic – portrait is among the surviving paintings that allegedly depict Falloppia or may at least have served as the model for later copies.

The most impressive portrait that claims to represent Falloppia is today in the library of the Botanical Garden in Padua; a digital image is accessible online.[37] It shows a scholar with a cap and a long beard, with an open book in front of him, in which lies a pressed and dried plant. Before him, we see a skull and next to his hand an elongated, slender metal instrument with rounded ends on both sides, by all appearances a surgical probe.

The inscription "GAB FALLOPIVS MVTINENSIS" is clearly legible on the upper left while another inscription "GABRIEL FALLOPIUS" on the upper right is faded and hardly visible anymore. According to Favaro, it bore the (erroneous) inscription "Gabriel Fallopius Mutinensis Lector Simplicium hortique curator ab anno MDLI ad MDLXIII". The catalogue of the holdings of the Botanical Garden assumes that it was painted in the nineteenth century but, of course, this does not exclude the possibility that it is a copy of a much earlier – or indeed contemporary and authentic – original.

At the beginning of his book on Falloppia, Favaro published a black-and-white reproduction of a portrait that was once owned by Giuseppe Sperino but by Favaro's time had been donated to the Anatomical Institute in Modena. From the black-and-white reproduction it would appear to be a copy of inferior artistic quality of the portrait in the Library of the Botanical Garden in Padua, showing Falloppia in exactly the same posture and in the same dress with a little piece of white color showing at the neck. The parts of the Padua painting with the hands, the book, the skull and the surgical probe are missing but they may very well have been cut off or may not have been copied in the first place.[38]

Figure 7.1 Alleged portrait of Falloppia in the Aula di Medicina of the Palazzo Bò in Padua (photo: Wellcome Collection, London).

Another, quite different portrait that claims to represent Falloppia hangs today on the walls of the Aula di Medicina in the main university building in Padua, the Palazzo Bò (see Figure 7.1). It shows a man with a blond beard and beret and carries the (faulty) inscription, referring to his teaching in Padua, "GABRIEL FALLOPIVS MVTINENSIS AN[ATOMICUS] AB ANNO 1551 AD ANNO 1563". Again there is not the slightest evidence that this is an authentic portrait.[39]

Falloppia's portrait can also be found in minor variations in printed works. The different versions probably have a common origin, a portrait by H. David, which was already published by Tomasini in 1630 (Figure 7.2) and may have served as a model for others. The portrait carries the inscription "GABRIEL FALLOPIVS MUTIN.[IS] PHILOSOPHVS ET MEDICVS" and shows the bearded Falloppia wearing a simple black cap and a robe or coat and holding a book in his right hand.[40]

Particularly fanciful is a portrait, which is supposed to represent Falloppia, that was printed in Budaeus' *Thanatologia* of 1707[41] and in the 1715 German edition of the *Secreti* falsely attributed to Falloppia[42] (Figure 7.3). It shows a man in his younger years, with a face that looks very different from

Figure 7.2 Alleged portrait of Falloppia in: Tomasini, Illustrium virorum (1630), p. 41.

Figure 7.3 Alleged portrait of Falloppia, in: Budaeus, Thanatologia (1707), inserted
      after p. 238 (also in: *Gabrielis Falopii [...] neu eröffnete vortreffliche rare
      Geheimnisse der Natur*, Frankfurt: Verlegt in Christian Genschen Buch-
      handlung 1715, before the frontispiece).

the one shown in Tomasini. In his right hand, he holds a long, slender glass
vessel, probably a chemical vessel, since the neck is far too thin for a urine
glass. The left hand rests on a nativity, which is lying in front of him. In the
two upper corners of the painting are two more nativity drawings, namely a
blank scheme with the twelve houses and a scheme with numbers and planetary and zodiacal signs and with a globe in the center – presumably a celestial globe. The caption reads "GABRIEL FALOPIUS CELEBERRIMUS
MEDICUS ET ASTROLOGUS IN VENET. ET PADVA AET. S. LXXIII."
Falloppia never reached the age of 73 and was very critical of astrology.
Probably the portrait offers a visual elaboration of Tomasini's faulty biographical sketch of Falloppia of 1630: Tomasini had claimed that Falloppia
was also a very knowledgeable astrologer and that he died at the age of 73.[43]

Notes

1 Stolberg, Learned physicians (2022); De Renzi, Career (2011).

2 See, for example, Coiter, Externarum (1573).
3 Carcano, Anatomici libri (1574).
4 Fabrizi's studies in Padua, in the 1550s, are poorly documented and we do not even know the year he obtained his doctoral degree. He clearly enjoyed an excellent training in surgery and anatomy, however, and surely Falloppia was the one person in Padua who could offer that training; see Favaro, Contributi (1922); Scipio, Girolamo Fabrici (1978); Rippa Bonati, Girolamo Fabrici d'Acquapendente (2004).
5 Archivio antico dell'Università di Padova, filza 665, professori di chirurgia e anatomia, teatro anatomico, ed altro, intorno a dette scuole, fol. 12v, ed. in Favaro, Gabrielle Falloppia (1928), p. 216, "1563 e 1564. Fu vacante la predetta lettura. Ma nel 1562 11 Xmbre fu eletto Francesco Ledinara per incisore con fiorini 35". Universitätsbibliothek Erlangen, Trew, Purkircher, Nr. 3, letter from Georg Purkircher to Joachim Camerarius, Padua, 27 January 1563, on Lendinara, Bellocati and Wieland.
6 Favaro, Contributi (1922); Scipio, Girolamo Fabrici (1978); on Fabrizi's teaching see Favaro, L'insegnamento (1922); Stolberg, Learning (2018).
7 On the influence of Galen's teleology on Vesalius see Siraisi, Vesalius (1997).
8 Andrew Cunningham has even described Fabrizi's anatomy as an "Aristotelean project" (Cunningham, Fabricius (1985); idem, Anatomical Renaissance (1995), pp. 171–182); I would agree with Nancy Siraisi (Siraisi, Historia (2004)), however, that Fabrizi combined Aristotelianism with Galenic teleology rather than supplanting one with the other.
9 On Casseri and his possible role in the history of these anatomical plates, see Stolberg, Teaching anatomy (2018).
10 Biblioteca Marciana, Venice, Rari 110–117; Rippa Bonati/ José Pardo-Tomàs, Teatro (2004).
11 Cunningham, Fabricius (1985); idem, Anatomical Renaissance (1995), pp. 172–173; Klestinec, Theaters (2011).
12 Stolberg, Learning (2018).
13 *Omnia, quae adhuc extant opera*. Venice: apud F. Valgrisium 1584.
14 *Opera quae adhuc extant omnia*. Frankfurt: apud haeredes Andreae Wecheli 1584.
15 *Opera omnia*. 2 vols. Frankfurt: apud haeredes Andreae Wechelii, Claud. Marnium & Io. Aubrium 1585.
16 Ibid., letter from Crato to Scutellarius, Breslau, 1 May 1585.
17 *Opera omnia*. 2 vols. Frankfurt: apud haeredes Andreae Wecheli, Claud. Marnium et Io. Aubrium 1600.
18 *Opera genuina omnia*. 3 vols. Venice: apud Io. Antonium & Iacobum de Franciscis 1606; ibid., vol. 1, letter to the reader.
19 Ibid., vol. 1, letter to the reader, "plane imperfectam, inordinatam, laceram, confusam, mutilatam".
20 *Opera omnia. Appendix*. Frankfurt: apud haeredes Andreae Wecheli, Claud. Marnium et Io. Aubrium 1606.
21 The complete title was *Secreti diversi et miracolosi ne' quali si mostra la via facile di risanare tutte le infirmità del corpo humano; et etiandio s'insegna il modo di fare molte altre cose, che à ciascuno sono veramente necessarie. Raccolti dal Eccel.mo Gabriele Falloppia; et da varie persone esperimentati*. Venice: Appresso Marco di Maria 1563.
22 Venice: Appresso Marco di Maria 1565; Venice: Appresso Vincenzo Valgrisi 1570; Venice: Appresso Alessandro Gardano 1578; Turin: Giovanni Varrone & Manfredo Morello 1580 (according to the catalogue of the Biblioteca civica in Cosenza; the entry is based on a damaged copy, however); Venice: Franceschini 1582 (according to the catalogue of the Universitätsbibliothek Wien); Venice: Ventura de Salvador 1585; Venice: Appresso Marc'Antonio Bonibelli

1597; Venice: presso Giov. Batt. Bonfadino 1611 (according to the catalogue of the Biblioteca cantonale, Lugano); Venice: Appresso Pietro Miloco 1620; Venice: Appresso Ghirardo Imberti 1640; Venice: Milochi 1640 (according to the catalogues of the Staats- und Stadtbibliothek Augsburg and of the Österreichische Nationalbibliothek in Vienna); Venice: Miloco; Venice: Conzatto 1676 (according to the catalogue of the Universitätsbibliothek Frankfurt).

23 Augsburg: Michael Manger, in Verlegung von Georg Willer 1571, with a dedicatory letter, dated 1 December 1570, to Archduke Ferdinand II by the translator, the Augsburg physician Jeremias Mertz. The Universitätsbibliothek in Eichstätt owns another Augsburg edition of 1571, by Willer, entitled "Secreta, oder heymliche Künst"; I have only seen the different frontispiece; presumably the two editions are otherwise identical; Augsburg: Michael Manger 1573 (according to the catalogue of the Staatsbibliothek Berlin); Augsburg: Michael Manger 1584 (according to the catalogue of the Staats- und Stadtbibliothek Augsburg); Augsburg: Michael Manger 1593; Augsburg: Michael Manger 1597; Frankfurt am Main: bey Nicolas Hoffmann/ In Verlegung Lucae Iennis 1616; Frankfurt am Main: Christoph le Blon 1641 (according to the catalogue of the Universitätsbibliothek Salzburg); Frankfurt am Main: Le Blon 1651 (according to the catalogue of the Landesbibliothek Coburg); Hamburg: Guht 1651 (according to the catalogue of the Universitätsbibliothek Erlangen). Hausmann, Bibliographie (1992/2018), pp. 459–462, provides information on some of the German editions and their printers and publishers.

24 *Fünfftzig außerwelte vnn heimliche stück des fürtrefflichsten Meisters Fallopia de Modena.* Nürnberg: Friderich Gutknecht sine anno. The printer Friderich Gutknecht is documented in the late 1570s and early 1580s, when he published various little astrological *practicae*, that is, astrological predictions for the following year.

25 Frankfurt am Main: Henning Grossen 1690 (according to catalogue of the Staats- und Stadtbibliothek Augsburg); Frankfurt am Main: Christian Gensch 1715.

26 For example, Staats- und Stadtbibliothek Augsburg, Med. 1150 (Secreti, Venice 1570) formerly owned by the monastery of the Franciscans; Bayerische Staatsbibliothek, Munich, 4° M.med. 78 (Secreten, Frankfurt 1616), formerly owned by the Order of Saint Jerome in Munich; Staatliche Bibliothek Regensburg, Med. 897 (Secreti, Venice 1597), formerly owned by the monastery of the Carmelites in Regensburg; ibid., 8° Med. 240 (Neu eröffnete [...] Geheimnisse, Frankfurt 1715), formerly owned by the monastery of the Franciscans in Regensburg.

27 *Gabrielis Falopii [...] neu eröffnete vortreffliche rare Geheimnisse der Natur.* Frankfurt: Verlegt in Christian Genschen Buchhandlung 1715.

28 Dedicatory letter to Bernardo Naugerio, 1 January 1564, in Falloppia, De medicatis aquis (1564).

29 Eamon, Science (1994); idem, How to read (2011).

30 *Secreti diversi* 1563, fol. 3 v.

31 Mazzuchelli (Mazzuchelli, Scrittori, vol. 2,3 (1762), p. 1531) and Tiraboschi (Tiraboschi, Biblioteca modenese (1782), p. 251) mistook the *Secreti* with its hundreds of recipes for Italian editions of the much smaller *Arcana* (see below).

32 Preface to *Arcanorum liber*, in: Agathus, Opuscula (1566), part 2, fol. 61r–v.

33 On Padovani see Dondi, Elideo Padovani (1951).

34 Falloppia, Opuscula (1566), fol. 62v.

35 Favaro, Gabrielle Falloppia (1928), pp. 155–157, mentions further engravings of dubious authenticity.

36 Wieland, Papyrus (1572), p. 100.
37 https://phaidra.cab.unipd.it/view/o:4704.
38 Favaro, Gabrielle Falloppia (1928), reproduced before the frontispiece.
39 See https://en.wikipedia.org/wiki/Gabriele_Falloppio (accessed 5 January 2022); also reproduced in Van Hee, Relationship (2017), p. 332.
40 Tomasini, Illustrium virorum (1630), p. 41.
41 Budaeus, Thanatologia (1707), inserted after p. 238.
42 Gabrielis Falopii […] neu eröffnete vortreffliche rare Geheimnisse der Natur, Frankfurt 1715, before the frontispiece.
43 Tomasini, Illustrium virorum (1630), pp. 42–43.

Sources and bibliography

Illustrations

Figure 2.1 Student's drawing showing the branching off of the humeral vein "according to Vesalius" (below) and "according to truth" (above), SUBG, Ms. Meibom 20, fol. 133r

Figure 4.1 Apparatus for the distillation of mineral waters, in: Falloppia, De aquis (1564), fol. 35v

Figure 7.1 Alleged portrait of Falloppia, in the Aula di Medicina in the Palazzo Bò in Padua (photo: Wellcome Collection, London)

Figure 7.2 Alleged portrait of Falloppia, in: Tomasini, Illustrium virorum (1630), p. 41.

Figure 7.3 Alleged portrait of Falloppia, in: Budaeus, Thanatologia (1707), inserted after p. 238 (also in: *Gabrielis Falopii [...] neu eröffnete vortreffliche rare Geheimnisse der Natur*, Frankfurt: Verlegt in Christian Genschen Buchhandlung 1715, before the frontispiece)

Manuscript sources

Basel, Universitätsbibliothek
Frey-Gryn Mscr. I 11, correspondence of Theodor Zwinger and others
Frey-Gryn Mscr. II 5, no. 9, *De cauterijs*[1]
Berlin, Staatsbibliothek
Sammlung Darmstaedter 3c 1550, letter from Gabrielle Falloppia to Giovanni Francesco Canani, Padua, 1 April 1561[2]
Sammlung Darmstaedter 3d 1546, undated letter from Antonio Fracanzano to Giovanni Francesco Canani in Ferrara
Sammlung Darmstaedter 3d 1553, letter from Agostino Gadaldini to Ludovico Castelvetro, Venice, 27 October 1553
Bologna, Biblioteca Universitaria
Mss Aldrovandi, 38[2], correspondence of Ulisse Aldrovandi, with letters from Falloppia,[3] Mattioli,[4] and Wieland[5]
Mss Aldrovandi, 98[1], with a list of plants in the botanical garden in Padua that Aldrovandi requested from Falloppia
Bologna, Archivio di Stato
Archivio del Reggimento, Litterarum Reg. n. 18 (7), 1561–1563, copy of letter from the Senate of Bologna to Falloppia, 1561[6]
Archivio del Reggimento, Assunteria di Studio, C n. 9, lettera F, tomo I, with a letter from Falloppia to the Senate of Bologna, Padua, 6 November 1561[7]

Dublin, Marsh's Library

Ms Z 4. 5. 2aa, 1576, Gabrielle Falloppia, "A ryght profitable and excellent treatyse teaching the true Methode [...] to cure [...] Morbus gallicus"[8]

Erlangen, Universitätsbibliothek

Ms. 908, student notes by an unidentified scribe, Helmstedt and Padua, early 1590s

Ms. 910, student notes by Johannes Brünsterer, Padua, early 1550s

Trew collection, letters from Georg Purkircher to Joachim Camerarius

Ferrara, Biblioteca comunale

Autografi, Racc. Cittadella, n. 1084, undated request by Falloppia that 8 lire from his salary be paid to the *bidello* of the university[9]

Florence, Archivio di Stato

Archivio Mediceo del Principato, Carteggio di Cosimo I, Filza 390, with a letter from Falloppia to the ducal secretary Giovan Lottini in Florence, 1548[10]

Archivio Guidi, Ms. 571, letters concerning the Studio Pisano, with two letters from Falloppia, Pisa, January and December 1550

Forlì, Biblioteca comunale Aurelio Saffi

Autografi Piancastelli 867, with a letter from Falloppia to Ulisse Aldrovandi, Padua, 4 December 1561[11]

Autografi Piancastelli 1431, with a letter from Pietro Andrea Mattioli to Ulisse Aldrovandi, 20 September 1559[12]

Göttingen, Niedersächsische Staats- und Universitätsbibliothek (SUBG)

Ms. Meibom 20, student notes on Falloppia's anatomical demonstrations, 1551–1553

Gotha, Forschungsbibliothek

Chart. A 629, student notes by Johann Mattenberg, Padua, 1570s

Kassel, Murhardsche Bibliothek

4 Ms. med. 19, foll. 1r–140r, student notes on Falloppia's lectures on ulcers and the French disease (1555)

London, British Library

Sloane Ms. 3293, English translation of Falloppia's, *De ulceribus* and *De tumoribus praeter naturam* by James Molins

Add. Ms. 10266, correspondence of Pietro Vettori, vol. 5, with a letter from Falloppia to Vettori

London, Wellcome Library

Western Manuscripts 269, student notes on Falloppia's lectures on ulcers and the French disease, 1560/61, with a manuscript copy of the funeral speech on Falloppia by Zamoscius[13]

Modena Biblioteca Estense

Mss ital. n. 833, letter from Falloppia to Luigi d'Este, Padua, 2 July 1556[14]

Modena, Archivio di Stato

Arch. per materie: Medici e medicine, Documenti esposti all'Esposizione Nazionale di Torino (1898), with four letters from Falloppia to Alfonso II d'Este, 1560[15]

Arch. per materie: Medici e medicine, b. 5, letters from Alfonso II d'Este to Falloppia, 11 March, 6 July, and 4 October 1562.[16]

Serie Particulari: Falloppia (Bartolomeo), letter from Alfonso II d'Este to Falloppia, 7 April 1560[17]

Naples, Biblioteca nazionale

Ms. V H 203, with notes or excerpts on the anatomy of the uterus, "Ex Falloppio" (foll. 8r–16v) and on a *Ratio habenda consultatione* "Ex Bassiano Lando placentino" (fol. 23r–v)

184   *Sources and bibliography*

Oxford, Bodleian Library

MS. Rawl. C. 724, with an English translation of Agathus' notes on Falloppia's *De morbo gallico* by the physician and surgeon Philipp Moore

MS. Canon. Misc. 115, student notes on Falloppia's lectures *De luxationibus* and *De fracturis*

MS. Canon. Misc. 119, student notes on 30 lectures by Falloppia *De aquis thermalibus* (1559)[18]

Paris, Bibliothèque nationale

Ms. Latin 13063, correspondence of Jacques Dalechamps

Padua, Archivio antico dell'Università di Padova

n° 675, letters to the Riformatori allo Studio, with two letters from the arts faculty on anatomical teaching December 1554 and January 1555[19]

Padua, Archivio di Stato

Archivio civico antico, Ducali 6, Registro

Archivio civico antico, Ducali 85, 1548–1556

Parma, Archivio di Stato

Epistolario scelto, b. 8, with a *consilium* by Falloppia and other professors in Padua for Ottavio Farnese, Duca di Parma, Padua, 9 May 1556[20]

Parma, Biblioteca Palatina

Autografi, Fondo di Lucca, casseta 2, letter from Falloppia to Lodovico Beccadelli, 1 October 1562[21]

Arch. Beccadelli, Ms. 1013, copy of a letter from Beccadelli to Falloppia, 6 October 1562[22]

Prague, Národní Muzeum, Knihovna

Nostická knihovna, H 365, Johannes Crato's personal copy of Mattioli, Epistolarum (1561), with a dedicatory letter from Mattioli, Prague, 8 February 1561, on the inside of the binding[23]

Rome, Biblioteca Vaticana

Reg. lat. 1297, pp. 191–193, copy of an epistolary consilium by Falloppia for *impotentia coeundi*, directly addressed to the (unidentified) patient

Siena, Biblioteca Comunale

Misc. XVI, C IX 17, anatomical notes (*syngrammata* and *antigrammata*) by Bartolommeo Eustachio

Misc. XVI, C IX 32, with student notes by an unidentified student on three *collegia* in which Falloppia participated

St. Petersburg, Российская Национальная Библиотека (Russian National Library)

Lat. F VI 95, lecture notes, among others on Falloppia, *De tumoribus*[24]

Urbania, Biblioteca comunale

Ms. 95, with notes of an unidentified student on Falloppia's lectures on the French disease (incomplete), ulcers of the eye, and injuries, 1561

Venice, Archivio di Stato (ASV)

Lettere dei Riformatori dello studio, filza 63 (1555–1559), with a letter (ibid., foll. 67r–68v) from Falloppia, Padua 12 December 1556[25]

Venice, Biblioteca Marciana

Rari 110–117, collection of colored anatomical illustrations

Ms. Ital. X, 142 (6407), *Monumenti spettanti allo studio di Padova*, with a letter (foll. 67r–68v) from Falloppia to the *Riformatori* in Venice, 21 July 1552[26]

Verona, Biblioteca civica

Ms. 169, handwritten summary of parts of Falloppia's work *De thermalibus aquis*[27]

Vienna, Österreichische Nationalbibliothek, Wien (ÖNBW)

Cod. 11200, collection of *experimenta* (proven remedies)

Cod. 11205, student notes by Georg Handsch on patient visits and medical practice

Cod. 11210, student notes by Georg Handsch, among others on Falloppia's anatomical demonstrations

Cod. 11224, student notes by Georg Handsch on Bassiano Landi's lectures

Cod. 11225, student notes by Georg Handsch on Falloppia's lectures on preternatural tumors (1552 and 1553)

Cod. 11251, collection of *experimenta* (proven remedies)

Wolfenbüttel, Herzog-August-Bibliothek

Cod. Guelf. 22 Aug. 4°, notes by an unidentified student on collegia and lectures by Falloppia in Padua (1555–1556)[28]

Extravagantes 264.2, notes by an unidentified student on Falloppia's lectures *De vulneribus*

Wrocław, Biblioteka Uniwersytecka

R 246, Nr. 104 (new: 102), with a letter from Joachim Camerarius to Johannes Crato von Krafftheim, 28 September 1561

Zürich, Zentralbibliothek

Ms. F 38, collection of letters written by Georg Keller

Bibliography of Falloppia's works

Collected works

Opuscula. Ed. by Petrus Angelus Agathus. Padua: apud Lucam Bertellum 1566 (*De morbo gallico, De bubone pestilenti, De principio venarum, De balsamo, De decoratione, De caloribus, Arcanorum liber*).

Opuscula tria. Venice: apud Paulum Meietum & Fratrem 1569 (*Tractatus de vulneribus, Expositio in lib. Hipp. de vulneribus capitis, Tractatus de vulneribus oculorum aliarumque partium corporis*.

Further edition: De parte medicinae, quae chyrurgia nuncupatur nec non in lib. Hippocratis de vulneribus capitis dilucidissima interpretatio. Venice: apud Paulum, & Antonium Fr. Meietos 1571.[29]

La chirurgia. Transl. of student notes on Falloppia's surgical lectures by Giovanni Pietro Maffei. Venice: Appresso Daniel Zanetti 1602.

Further editions:

Venice: Presso Giacomo Anton. Somascho 1603

Venice: apresso Vincenzo Somascho 1620

Venice: Appresso Pietro Maria Bertano 1637[30]

Venice: appresso i Bertani 1647

Venice: presso Steffano Curti 1675

Venice: presso Paolo Baglioni 1675

Venice: presso Abondio Menafoglio 1675[31]

Omnia quae adhuc extant opera. Venice: apud Felicem Valgrisium 1584.

Opera quae adhuc extant omnia. Frankfurt: apud haeredes Andreae Wecheli 1584.

Opera omnia. 2 vols. Frankfurt: apud haeredes Andreae Wechelii, Claud. Marnium & Io. Aubrium 1585.

Opera omnia. 2 vols. Frankfurt: apud haeredes Andreae Wecheli, Claud. Marnium et Io. Aubrium 1600.

Appendix. Frankfurt: Typis Wechelianis apud Claud. Marnium et heredes Ioan. Aubrii 1606 (additional material, taken from the Venice edition of 1606).

Opera genuina omnia. 3 vols. Venice: apud Io. Antonium & Iacobum de Franciscis 1606.

Anatomy

Observationes anatomicae

Observationes anatomicae ad Petrum Mannam. Venice: apud Marcum Antonium Vlmum 1561.

Further editions:

Venice: apud Marcum Antonium Vlmum 1562.

Cologne: apud haeredes A. Birckmanni 1562.

Paris: apud Iacobum Keruer 1562.

In: Opera (Venice) 1584, foll. 221r–266v.

In: Opera (Frankfurt) 1584, pp. 398–482.

In: Opera 1585, vol. 1, pp. 354–430.

Helmstedt: Lucius 1588, ed. by Johannes Sigfridus.

In: Opera 1600, vol. 1, pp. 354–430.

In: Opera 1606, vol. 1, pp. 37–115.

In: Andreas Vesalius: Opera omnia anatomica et chirurgica. Leiden: apud J. du Vivie and Ioan. and Herm. Verbeek 1725, pp. 687–758.

Modena: S.T.E.M. Mucchi 1964, photographic reproduction of the 1561 edition with an Italian translation by Gabriella Righi Riva and Pericle Di Pietro in vol. 2.

De principio venarum

Quaestio de principio venarum: In: *Opuscula*. Ed. by Petrus Angelus Agathus. Padua: apud Lucam Bertellum 1566, part 2, foll. 17r–23r.

De ossibus

Expositio in librum Galeni de ossibus. Ed. by Franciscus Michinus de Sancto Arcangelo. Venice: apud Simonem Galignanum de Karera 1570, foll. 1r–70v.

Further editions:

In: Opera (Venice) 1584, foll. 289–330v

In: Opera (Frankfurt) 1584, pp. 520–595

In: Opera 1585, vol. 1, pp. 464–530

In: Opera 1600, vol. 1, pp. 464–530

In: Opera 1606, vol. 3, foll. 121r–156v

Observationes anathomicae / Observationes de venis

Observationes anathomicae. In: De ossibus (1570), foll. 71r–76r.

Further editions (as *Observationes de venis*):

In: Opera (Venice) 1584, foll. 331r–333v

In: Opera (Frankfurt) 1584, pp. 506–600

In: Opera 1585, vol. 1, pp. 531–534

In: Opera 1600, vol. 1, pp. 531–534

In: Opera 1606, vol. 1, pp. 116–120

De humani corporis anatome compendium / Institutiones anatomicae

De humani corporis anatome compendium. Venice: apud Paulum & Antonium Meietos fratres 1571.

Further editions:

In: Opera (Venice) 1584, foll. 267–288v

In: Opera (Frankfurt) 1584, pp. 482–520
Padua: apud Paulum Meietum 1585
In: Opera 1585, vol. 1, pp. 430–464
In: Opera 1600, vol. 1, pp. 430–464
In: Opera 1606, vol. 1, pp. 1–36

De partibus similaribus

De partibus similaribus humani corporis. Ed. by Volcher Coiter. Nürnberg: in
  officina Theodorici Gerlachii 1575
Further editions
In: Opera 1585, vol. 2, pp. 96–156
In: Opera 1600, vol. 2, pp. 96–156
In: Opera 1606, vol. 1, pp. 121–184

Surgery

De bubone pestilenti

Tractatus de bubone pestilenti. In: *Opuscula.* Ed. by Petrus Angelus Agathus. Padua:
  apud Lucam Bertellum 1566 (part 2, foll. 1r–16v).[32]

De cauteriis

De cauteriis. In: *De compositione medicamentorum* (1570), foll. 65r–72v.
Further editions:
In: Opera (Venice) 1584, foll. 334–340v
In: Opera (Frankfurt) 1584, pp. 600–611
In: Opera 1585, vol. 1, pp. 534–544
In: Opera 1600, vol. 1, pp. 534–544
In: Opera 1606, vol. 2, pp. 457–467

De decoratione

Tractatus de decoratione. In: *Opuscula.* Ed. by Petrus Angelus Agathus. Padua:
  apud Lucam Bertellum 1566, part 2, foll. 34r–51v.
Further editions
In: Opera 1585, vol. 2, pp. 325–344
In: Opera 1600, vol. 2, pp. 325–344
In: Opera 1606, vol. 3, foll. 110r–120v

De fracturis

De fracturis. In: Opera 1585, vol. 2, pp. 88–95
Further editions:
In: Opera 1600, vol. 2, pp. 88–95
In: Opera 1606, vol. 3, foll. 184v–191v (as part of *De luxatis et fractis ossibus*)

De luxationibus

De luxationibus. In: Opera 1600, vol. 2, pp. 60–87
Further editions: *De luxationibus.* In: Opera 1600, vol. 2, pp. 60–87
In: Opera 1606, vol. 3, foll. 157r–184v (as part of *De luxatis et fractis ossibus*)

De luxatis et fractis ossibus

De luxatis et fractis ossibus. In: Opera 1606, vol. 3, foll. 157r–191v

De morbo gallico

De morbo gallico liber absolutissimus. Ed. by Petrus Angelus Agathus. Padua: apud
  Lucam Bertellum & socios 1563.
Further editions:

Venice: Ex officina Francisci Laurentini de Turino 1565
In: Opuscula 1566 (separate pagination)
Venice: apud Haeredes Melchioris Sessae 1574
Venice: ex officina Francisci Laurentini de Turino 1574
Venice: apud Aegidium Regazolam 1574
In: Opera (Venice) 1584, foll. 427–469v
In: Opera (Frankfurt) 1584, pp. 770–848
In: Opera 1585, vol. 1, pp. 682–749
In: Opera 1600, vol. 1, pp. 682–749
In: Opera 1606, vol. 2, pp. 113–203 (as part of *De ulceribus*)

De tumoribus praeter naturam

De tumoribus praeter naturam. In: *Libelli duo alter de ulceribus, alter de tumoribus.*
   Venice: apud Donatum Bertellum 1563, foll. 33r–101r.

Further editions:

In: Libelli duo, Venice: apud Donatum Bertellum 1566, foll. 33r–102r
In: Opera (Venice) 1584, foll. 387–426v
In: Opera (Frankfurt) 1584, pp. 696–770
In: Opera 1585, pp. 618–682 and vol. 2, pp. 253–255, as *Tractatiuncula sive potius
   lectio de tumore bilioso, quam benignus lector ad doctrinae tomi primi de tumoribus
   praeter naturam caput 28. quod de erysipelate est, referre potest*, and pp. 256–324,
   *De scirrho et reliquis tumoribus, cum universalibus tum particularibus, qui libro ei-
   usdem in tomum primum relato desunt.*
In: Opera 1600, vol. 1, pp. 618–682 and, vol. 2, pp. 253–255, as *Tractatiuncula sive
   potius lectio de tumore bilioso, quam benignus lector ad doctrinae tomi primi de
   tumoribus praeter naturam caput 28. quod de erysipelate est, referre potest*, and
   pp. 256–324, *De scirrho et reliquis tumoribus, cum universalibus tum particularibus,
   qui libro eiusdem in tomum primum relato desunt.*
In: Opera 1606, vol. 3, foll. 1r–109v

De ulceribus

De ulceribus. In: *Libelli duo alter de ulceribus, alter de tumoribus.* Venice: apud
   Donatum Bertellum 1563, foll. 1r–32v.

Further editions:

In: *Libelli duo*, Venice: apud Donatum Bertellum 1566, foll. [0]r^{33}-32r
De ulceribus liber. Ed. Bruno Seidel. Erfurt: Georgius Baumanus 1577
In: Opera (Venice) 1584, foll. 368–386v
In: Opera (Frankfurt) 1584, pp. 661–696
In: Opera 1585, vol. 1, pp. 587–618 and vol. 2, pp. 236–253, *De ulceribus particularibus*
In: Opera 1600, vol. 1, pp. 587–618 and vol. 2, pp. 236–253, *De ulceribus particularibus*
In: Opera 1606, vol. 2, pp. 1–235

De vulneribus

Tractatus de vulneribus. In: Opuscula tria. Venice: apud Paulum Meietum & Fratrem 1569, pp. 1–94.

Further editions:

In: Opera 1585, vol. 2, pp. 157–195 (*De vulneribus in genere seu universali*)
In: Opera 1600, vol. 2, pp. 157–195 (*De vulneribus in genere seu universali*)
In: Opera 1606, vol. 2, pp. 236–313

De vulneribus capitis

In Hippocratis librum de vulneribus capitis expositio. Ed. by Petrus Angelus Agathus.
   Venice: apud Lucam Bertellum 1566.

Further editions:

In: Opuscula tria. Venice: apud Paulum Meietum & Fratrem 1569, pp. 101–180.

In: Opera (Venice) 1584, foll. 341–367v

In: Opera (Frankfurt) 1584, pp. 612–660

In: Opera 1585, vol. 1, pp. 545–587

In: Opera 1600, vol. 1, pp. 545–587.

In: Opera 1606, vol. 2, pp. 411–456

De vulneribus oculorum aliarumque partium corporis

Tractatus de vulneribus oculorum, aliarumque partium corporis. Venice: apud
Paulum Meietum & Fratrem 1569, pp. 183–223.

Further editions:

In: Opera 1585, vol. 2, pp. 196–236. *De vulneribus particularibus a capite ad infra versus*

In: Opera 1600, vol. 2, pp. 196–236. *De vulneribus particularibus a capite ad infra versus*

In: Opera 1606, vol. 2, pp. 314–410 (as part of *De vulneribus*)

Materia medica

De compositione medicamentorum

De compositione medicamentorum. Venice: apud Paulum & Antonium Meietos
fratres (ex officina Gratiosi Perchacini) 1570.

Further editions:

In: Opera (Venice) 1584, foll. 79–117v

In: Opera (Frankfurt) 1584, pp. 144–215

In: Opera 1585, vol. 1, pp. 130–193

In: Opera 1600, vol. 1, pp. 130–193

In: Opera 1606, vol. 3, foll. 195r–228v

De materia medicinali

De materia medicinali in lib. I Dioscoridis. In: Opera omnia, vol. 2. Frankfurt:
apud haeredes Andreae Wechelii, Claud. Marnium & Io. Aubrium 1585,
pp. 25–59.

Further editions:

Tractatus de balsamo: In: Opuscula. Ed. by Petrus Angelus Agathus. Padua: apud
Lucam Bertellum 1566, part 2, foll. 23v–33v (fragment, limited to a few chapters
on *balsamum* and a handful of other substances)

In: Opera 1600, vol. 2, pp. 25–59.

In: Opera 1606, vol. 1, pp. 211–248.

De medicamentis simplicibus

*De medicamentis simplicib*us. In: Opera 1585, vol. 2, pp. 1–24.

Further editions:

In: Opera 1600, vol. 2, pp. 1–24.

In: Opera 1606, vol. 1, pp. 185–210

De medicatis aquis

De medicatis aquis atque de fossilibus. Ed. by Andrea Marcolino. Venice: apud
Ludovicum Avantium 1564, foll. 1r–85v.

Further editions:

Venice: apud Ludouicum Auantium 1564[34]

Venice: Ex officina Ludovici Avantii 1569

In: Opera (Venice) 1584, foll. 118–166v

In: Opera (Frankfurt) 1584, pp. 215–302
In: Opera 1585, vol. 1, pp. 193–269.
In: Opera 1600, vol. 1, pp. 193–269
In: Opera 1606, vol. 1, pp. 249–329

De metallis et fossilibus

De medicatis aquis atque de fossilibus. Ed. by Andrea Marcolino. Venice: apud
Ludovicum Avantium 1564, foll. 86r-76r

Further editions

In: Opera (Venice) 1584, foll. 167–220v
In: Opera (Frankfurt) 1584, pp. 302–397
In: Opera 1585, vol. 1, pp. 270–353
In: Opera 1600, vol. 1, pp. 270–353
In: Opera 1606, vol. 1, pp. 330–416

De simplicibus medicamentis purgantibus

De simplicibus medicamentis purgantibus. Ed. by Andrea Marcolino. Venice: Ad
insignem stellae Iordani Ziletti 1565.

Further editions:

Venice: Ad insignem stellae Iordani Ziletti 1566
In: Opera (Venice) 1584, foll. 1r–75v
In: Opera (Frankfurt) 1584, pp. 1–138
In: Opera 1585, vol. 1, pp. 1–124
In: Opera 1600, vol. 1, pp. 1–124.
In: Opera 1606, vol. 1, pp. 417–547

De asparagis

De asparagis (letter to Girolamo Mercuriale, 1 November 1558), in: *De simplicibus
medicamentis purgantibus tractatus* (1565), pp. 254–263.

Further editions:

In: Opera (Venice) 1584, foll. 76r–78v[35]
In: Opera (Frankfurt) 1584, pp. 139–144
In: Opera 1585, vol. 1, pp. 124–129
In: Opera 1600, vol. 1, pp. 124–129
In: Opera 1606, vol. 1, pp. 548–552

De caloribus

Tractatus de caloribus, in quo agitur de concoctione. In: Opuscula. Ed. by Petrus
Angelus Agathus. Padua: apud Lucam Bertellum 1566, part 1, foll. 56r–60v.

Further editions

Opuscula. Padua: apud Lucam Bertellum 1569

De modo consultandi

De modo consultandi. In: Opera 1606, vol. 3, foll. 192r–194r

Further editions:

In: Appendix 1606, pp. 94–98.

De obstructionibus

*De obstructionibus ingentibus circa mesenterium, a materia varia factis, D. Philippo Ban-
zollae Parmensi.* In: Girolamo Giunti (ed.): *De balneo thermali, Lixignano vocato,
necnon de luto barboliorum medicato, in ducatu Parmensi, tractatus Hieronymi Zun-
thi, philosophi, ac medici Parmensis. In quo breuiter docentur modi reales vtendi, &
aquis, & luto thermali.* Venice: apud hæredem Damiani Zenarij 1615, S. 99–102.

General bibliography

Achillini, Alessandro: *Annotationes anatomicae.* Bologna: H. de Benedictis 1520.

Agathus, Petrus Angelus (ed.): Arcanorum liber. In: idem (ed.): *Opuscula,* part 2. Padua: apud Lucam Bertellum 1566, foll. 61r–69v.

Alessandrini, Giulio: *De medicina et medico dialogus.* Zürich: per Andream Gesnerum 1557.

Amatus Lusitanum: *Curationum medicinalium centuria prima, multiplici variaque rerum cognitione referta.* Florence: Torrentinus 1551.

Ambrose, Charles T.: Immunology's first priority dispute. An account of the 17th-century Rudbeck–Bartholin feud. In: *Cellular Immunology* 242 (2006), pp. 1–8.

Anastasio, Pamela: Fracanzani, Antonio. In: *Dizionario Biografico degli Italiani* 49 (1997). http://www.treccani.it/enciclopedia/antonio-fracanzani (Dizionario-Biografico)/.

Andreozzi, Alfonso: *Le leggi penali degli antichi Cinesi: discorso proemiale sul diritto e sui limiti del punire e traduzioni originali dal cinese.* Florence: Giuseppe Civelli 1878.

Angelini, Alberto: *Una lettere inedita di Gabriele Fallopio.* Presented to the bride-groom on occasion of the wedding between Bice Visconti and Dott. Ricciotto Gozzini in Florence, 26 March 1900. Florence: Tipografia cooperativa 1900 (reprinted as *Gabriele Falloppio. Cenni biografici ed una lettera inedita* in *Supplementa al Policlinico* 6 (1900), pp. 1200–1203).

Anonymus: *Memorie per servire all'istoria letteraria.* Vol. 8, part III. Venice: Appresso Pietro Valvasene 1756.

Aranzi, Giulio Cesare: Anatomicae observationes. In: *De humano foetu liber, tertio editus, ac recognius. Eiusdem anatomicarum observationum liber ac de tumoribus secundum locos affectos liber nunc primum editi.* Venice: apud Iacobum Brechtanum 1587, pp. 41–119.

Argenterio, Giovanni: *De consultationibus medicis sive (ut vulgus vocat) de collegiandi ratione liber.* Florence: Cudebat Laurentius Torrentinus 1551.

Arrizabalaga, Jon, John Henderson and Roger French: *The great pox. The French disease in Renaissance Europe.* New Haven, CT: Yale University Press 1997.

Aselli, Gaspare: *De lactibus siue lacteis venis quarto vasorum mesaraicorum genere nouo inuento.* Milan: apud Io. Bapt. Bidellium 1627.

Astruc, Jean: *De morbis venereis libri novem.* Paris: Cavelier 1740.

Barotti, Lorenzo: *Memorie istoriche di letterati ferraresi.* Vol. 2. Ferrara: Rinaldi 1793.

Bartholin, Thomas: *Bartholinus anatomy.* Ed. by Nicholas Culpeper and Abdiah Cole. London: John Streater 1668.

Belloni Speciale, Gabriella: Falloppia, Gabriele. In: *Dizionario biografico degli italiani* 44(1994).https://www.treccani.it/enciclopedia/gabriele-falloppia_%28Dizionario-Biografico%29/.

Beloch, Giulio: Ricerche sulla storia della popolazione di Modena e del Modenese. In: *Rivista italiana di sociologia»* 12 (1908), pp. 1–48.

Belon, Pierre: *L'histoire de la natvre des oyseavx.* Paris: Cavellat 1555.

Berengario da Carpi, Jacopo: *Anatomia Carpi. Isagoge breves perlucide ac uberrime in anatomiam humani corporis.* Venice: per Bernardinum de Vitalibus 1535.

Biesbrouck, Maurits, Theodoor Goddeeris and Steno Steeno: Jean Tagault (c. 1486–1546), professor heelkunde in Parijs, plagiator van Vesalius' Tabulae anatomicae sex (1538)?. In: *In Monte Artium. Journal of the Royal Library of Belgium* 10 (2017), pp. 7–63.

Blumenbach, Johann Friedrich: Nachricht von der auf der Göttingischen Bibliothek befindlichen Meibomischen Sammlung medicinischer Handschriften. In: *Medicinische Bibliothek* 1 (1783), pp. 368–377.

Borsetti, Ferrante: *Historia almi Ferrariae Gymnasii*. 2 vols. Ferrara: Typis Bernardini Pomatelli 1735.

Brasavola, Antonio Musa: *Index in omnes Galeni libros*. Venice: apud Iuntas 1609.

Budaeus, Guilielmus: Thanatologia. In: Heinrich Turck (ed.): *Rerum germanicarum tres selecti scriptores*. Part 3. Frankfurt am Main: Ex officina Genschiana 1707, pp. 177–272.

Bylebyl, Jerome J.: Cardiovascular physiology in the sixteenth and early seventeenth centuries. Unpublished PhD Thesis. New Haven, CT: Yale University 1969.

Calderato, Vincenzo: *Brevi cenni sulla vita e sugli scritti anatomici di Gabriele Falloppio*. Diss. med. Padua: Pietro Prosperini 1862.

Campori, Giuseppe: Lettere inedite di Gabriello Falloppia e documenti relativi al medesimo. In: *Atti e memorie della R. Deputazione di Storia Patria per le Provincie Modenesi e Parmensi* 2 (1864), pp. 433–441.

Canalis, Rinaldo F.: Gabrielle Falloppia: Vesalius's admirer and first critic. In: idem and Massimo Ciavolella, Andreas Vesalius (2018), pp. 171–200.

Canalis, Rinaldo F.: Vesalius's methods in the production of the *Fabrica* with emphasis on the neuroanatomy chapters. In: idem and Massimo Ciavolella, Andreas Vesalius (2018), pp. 131–168.

Canalis, Rinaldo F. and Massimo Ciavolella (eds.): *Andreas Vesalius and the "Fabrica" in the age of printing: Art, anatomy, and printing in the Italian Renaissance*. Turnhout: Brepols 2018.

Canani, Giovanni Battista: *Musculorum humani corporis picturata dissectio*. [Ferrara? ca 1541]; modern edn by Giulio E. Muratori (=Archivio italiano di anatomia e di embriologia, vol. 67, n.1). Florence: Ed. Il Sedicesimo 1962.

Capivaccio, Girolamo: De medica consultandi ratione, seu de arte collegiandi, liber. In: idem: *Opera omnia*. Ed. by Johann Hartmann Beyer. Frankfurt: E Palthenia, curante Iona Rhodio 1603, pp. 931–940.

Capparoni, Pietro: *Profili bio-bibliografici di medici e naturalisti celebri italiani dal sec. XV° al sec. XVIII°*. Vol. 2. Rome: Istituto Naz. Medico-Farmacologico "Serono" 1928.

Carcano Leone, Giovanni Battista: *Anatomici libri II*. Pavia: apud Hieronymum Bartholum 1574.

Carcano Leone, Giovanni Battista: Liber II. In quo de musculis palpebrarum atque oculorum motibus deservientibus pertractatur. In: idem, *Anatomici libri II* (1574) (separate pagination).

Cardano, Girolamo: *De subtilitate libri XXI*. Basel: apud Sebastianum Henricpetri 1582.

Carlino, Andrea: *Books of the body. Anatomical ritual and Renaissance learning*. Chicago, IL: University of Chicago Press 1999.

Casoli, Vincenzo: *I sifilografi modenesi del sec. XVI (A. Fontana, A. Scanaroli, N. Macchelli, G. Falloppi)*. Modena: Tip. Lit. Bassi e Debri 1905.

Casoli, Vincenzo: Gli statuti del Collegio dei medici della città di Modena riformati da Giovanni Grillenzoni medico modenese (1501–1551). In: *Rivista di storia critica delle scienze mediche e naturali. Organo ufficiale della Società italiana di storia critica delle scienze mediche e naturali* 2 (1911), pp. 57–80 and 93–107.

Castellani, Aloysius Franciscus: *De vita Antonii Musae Brasavolae commentarius historico-criticus ex ipsius operibus erutus.* Mantua: Joseph Barglia 1767.

Castiglioni, Arturo: Fallopius and Vesalius. In: Harvey Cushing (ed.): *A biobibliography of Andreas Vesalius.* 2nd edn. Hamden, CT: Archon 1962, pp. 182–195.

Chauliac, Guy de: *Inventarium sive chirurgia magna.* Ed. by Michael R. McVaugh. Vol. 1: *Text.* Leiden: Brill 1997.

Coiter, Volcher: *Externarum et internarum principalium humani corporis partium tabulae, atque anatomicae exercitationes observationesque variae.* Nürnberg: in officina Theodorici Gerlatzeni 1573.

Cole, Francis Joseph: *A history of comparative anatomy. From Aristotle to the eighteenth century.* London: Macmillan & Co 1949.

Colombo, Realdo: *De re anatomica libri XV.* Venice: Bevilacqua 1559.

Cook, Harold J.: Victories for empiricism, failures for theory. Medicine and science in the seventeenth-century. In: Charles T. Wolfe and Ofer Gal (eds.): *The body as object and instrument of knowledge. Embodied empiricism and early modern science.* Dordrecht et al.: Springer 2010, pp. 9–32.

Cordus, Valerius: *Pharmacorum omnium, quae quidem in usu sunt, conficiendorum ratio. Vulgo vocant dispensatorium pharmacopolarum.* Nürnberg: Petreius 1546.

Corradi, Alfonso: *Tre lettere d'illustri anatomici del Cinquecento: Aranzio, Canano, Falloppia* (= extract from Annali universali di medicina 265 (1883)). Milan: Fratelli Rechiedei 1883.

Corradi, Alfonso: Degli esperimenti tossicologici in anima nobili nel '500. In: *Rendiconti dell'Istituto Lombardo di Science e Lettere* 19 (1886), pp. 361–363.

Costa, Emilio: *Ulisse Aldrovandi e lo studio bolognese nella seconda metà del secolo XVI. Discorso.* Bologna: Stabilimento poligrafico emiliano 1907.

Cunningham, Andrew: Fabricius and the "Aristotele project" in anatomical teaching and research at Padua. In: Andrew Wear, Roger French and Iain Lonie (eds.): *The medical renaissance of the sixteenth century.* Cambridge: Cambridge University Press 1985, pp. 195–222.

Cunningham, Andrew: *The anatomical renaissance. The resurrection of the anatomical projects of the ancients.* Aldershot: Scolar Press 1997.

Daza Chacón, Dionisio: *Pratica y teorica de cirugia en romance y en latin.* Valencia: por Francisco Cipres 1673.

De Caro, Raffaele: Anatomy in colour: Girolamo Fabrici d'Acquapendente (c. 1537–1619). In: Canalis and Ciavolella, Andreas Vesalius (2018), pp. 201–224.

De Renzi, Silvia: A career in manuscripts: Genres and purposes of a physician's writing in Rome, 1600–1630. In: *Italian Studies* 66 (2011), pp. 234–248.

De Toni, Giovanni Battista: *Spigolature aldrovandiane IX. Intorno alle relazioni del botanico Melchiorre Guilandino con Ulisse Aldrovandi.* In: Atti dell'I.R. Accademia di Scienze, Lettere ed Arti degli Agiati in Rovereto 17 (1911), pp. 149–171.

De Toni, Giovanni Battista: *Spigolature aldrovandiane X. Alcune lettere di Gabriele Falloppia ad Ulisse Aldrovandi.* In: Atti e Memorie della R. Deputazione di Storia Patria per le Provincie Modenesi, serie V, vol. VII (1913), pp. 34–46.

De Toni, Giovanni Battista: Bartolomeo Maranta. In: Mieli, Gli scienziati (1921), pp. 68–70.

De Toni, Giovanni Battista: Melchiorre Guilandino. In: Mieli, Gli scienziati (1921), pp. 73–76.

De Toni, Giovanni Battista: Luigi Anguillara. In: Mieli, Gli scienziati (1921), pp. 76–78.

Dear, Peter: The meanings of experience. In: Katharine Park and Lorraine Daston (eds.): *Early modern science.* Cambridge: Cambridge University Press 2006, pp. 106–131.

Debus, Allen G.: *The chemical philosophy. Paracelsian science and medicine in the sixteenth and seventeenth centuries.* Vol. 1. New York: Science History Publications 1977.

Delisle, Candice: The letter: private of public place? The Mattioli-Gesner Controversy about the aconitum primum. In: *Gesnerus* 61 (2004), pp. 161–176.

Di Pietro, Pericle: Nel quarto centenario della morte: Gabrielle Falloppia (1523–1562) nella storia dell'anatomia. In: *Bollettino della Società medico-chirurgica di Modena* 62 (1962), pp. 414–423.

Di Pietro, Pericle (ed.): *Epistolario di Gabriele Falloppia.* Ferrara: Università degli studi di Ferrara 1970.

Di Pietro, Pericle: Contributo alla biografia di Gabriele Falloppia. In: *Atti e memorie della Reale Accademia Petrarca di lettere, arti e scienze* 10 (1974), pp. 63–72.

Di Pietro, Pericle and Gabriella Cavazzuti: La descrizione falloppiana delle tube uterine. In: *Acta medicae historiae Patavina* 11 (1964), pp. 51–60.

Dioscorides, Pedanius: *De medica materia.* Lyon: Frellonius 1547.

Dondi, Giovanni de: De fontibus calidis agri Patauini consideratio. In: Giunta, De balneis (1553), foll. 94r–108v.

Dondi, Raffaele Flaminio: Elideo Padovani. Medico forlivese del secolo XVI. In: *Atti e memorie dell'Accademia di storia dell'arte sanitaria* 117 (1951), pp. 139–144.

Dubois, Jacques: *In Hippocratis & Galeni physiologiae partem anatomicam isagoge.* 2nd edn. Venice: Valgrisius 1556.

Du Chastel, Pierre: *Vitae illustrium medicorum qui toto orbe ad haec usque tempora floruerunt.* Antwerp: Apud Guilielmum a Tongris 1618.

Eamon, William: *Science and the secrets of nature. Books of secrets in medieval and early modern culture.* Princeton, NJ: Princeton University Press 1994.

Eamon, William: How to read a book of secrets. In: Elaine Leong and Alisha Rankin (eds.): *Secrets and knowledge in medicine and science, 1500–1800.* Farnham, Surrey and Burlington, VT: Ashgate 2011, pp. 23–46.

Eloy, Nicolas F. J.: *Dictionnaire historique de la médecine ancienne et moderne.* Vol. 2. Mons: Chez H. Hoyois 1778, pp. 193–196.

Erastus, Thomas: *De medicina nova Philippi Paracelsi.* Part 2. Basel: Perna 1572.

Eriksson, Ruben: *Andreas Vesalius' first public anatomy at Bologna 1540. An eyewitness report by Baldasar Heseler together with notes on Matthaeus Curtius' lectures on 'anatomia Mundini'.* Uppsala: Almqvist & Wiksell 1959.

Estienne, Charles: *De dissectione partium corporis humani libri III.* Paris: apud Simonem Colinaeum 1545.

Eustachius, Bartholomaeus: *Tabulae anatomicae.* Ed. by Giovanni Maria Lancisi. Rome: Ex Officina Typographica Francisci Gonzagae 1714.

Fabroni, Angelo: *Historia academiae pisanae.* Pisa: Cajetanus Mugnaius 1792.

Fabrucci, Stefano Maria: De pisano gymnasio sub Cosmo primo mediceo feliciter renovato. In: Angelo Calogerà (ed.): *Nuova raccolta d'opuscoli scientifici e filologici.* Vol. 6. Venice: Presso Simone Occhi 1760, pp. 1–137.

Fantuzzi, Giovanni: *Memorie della vita di Ulisse Aldrovandi, medico e filosofo Bolognese.* Bologna: Volpe 1774.

Favaro, Antonio (ed.): *Atti della nazione germanica artista nello Studio di Padova.* Vol. 1. Venice: A spese della Società 1911.

Favaro, Giuseppe: Contributi alle biografia di Girolamo Fabrici d'Acquapendente. In: *Memorie e documenti per la storia della Università di Padova.* Vol. 1. Padua: La Garangola 1922, pp. 241–348.

Favaro, Giuseppe: *Gabrielle Falloppia modenese (MDXXII-MDLXII). Studio biografico.* Modena: Tipografia editrice immacolata concezione 1928.

Ferinarius, Ioannes: Narratio historica de vita et morte viri summi Ioach[imi] Curei. Lignitz: Nicolaus Sartorius 1601.

Ferrari, Giorgio E.: Le opere a stampa del Guilandino. Per un paragrafo dell'editoria scientifica padovana del pieno Cinquecento. In: Antonio Barzon (ed.): *Libri e stampatori in Padova: Miscellanea di studi storici in onore di Mons. G. Bellini.* Padua: Tipografia Antoniana 1959, pp. 377–463.

Ferrari, G. E.: L'opera idro-termale di Gabriele Falloppio. Le sue edizioni e la sua fortuna. In: *Quaderni per la storia dell'Università di Padova* 18 (1985), pp. 1–41.

Ferretto, Silvia: *Maestri per il metodo di trattar le cose. Bassiano Lando, Giovan Battista da Monte e la scienza della medicina nel XVI secolo.* Padua: CLEUP 2012.

Firpo, Massimo and Dario Marcatto: *Il processo inquisitoriale del cardinal Giovanni Morone. Edizione critica.* Vol. 2: *Il processo d'accusa.* Rome: Istituto storico italiano per l'età moderna e contemporanea 1984.

Fortuna, Stefania: The Latin editions of Galen's *Opera omnia* (1490–1625) and their prefaces. In: *Early Science and Medicine* 17 (2012), pp. 391–412.

Foucard, Cesare: *Documenti storici spettanti alla medicina, chirurgia, farmaceutica conservati nell'Archivio di Stato in Modena.* Modena: Tip. sociale 1885.

Fracanzano, Antonio: *De morbo gallico fragmenta quaedam elegantissima, ex lectionibus anni MDLXII Bononiae.* In: Falloppia, De morbo gallico (1574), pp. 186–219.

Fracastoro, Girolamo: *Syphilis sive morbus gallicus.* Basel: Bebel 1536.

Franceschini, Pietro: Luci e ombre nella storia delle trombi di Falloppia. In: *Physis* 7 (1965), pp. 215–250.

Franklin, K. J.: Valves in veins. An historical survey. In: *Journal of the Royal Society of Medicine* 21 (1927), pp. 1–33.

Frigimelica, Francesco: *De balneis metallicis artificio parandis liber.* Nürnberg: Impensis Johannis Ziegeri 1679.

Gadebusch Bondio, Mariacarla: *Medizinische Ästhetik. Kosmetik und plastische Chirurgie zwischen Antike und früher Neuzeit.* Munich: Fink 2005.

Galen: *Opera omnia.* Venice: apud haeredes Lucae Antonij Iuntae 1541.

Galen: Thrasybulus. In: idem: *Hygiene. Books 5–6. Thrasybulus. On exercise with a small ball.* Ed. and trans. by Ian Johnston. Cambridge, MA/London: Harvard University Press 2018, pp. 224–371.

Gessner, Konrad: Euonymus. Vol. 2. Ed. by Caspar Wolff. [Zürich 1569] (date according to the dedicatory letter from Gessner, Zürich, 13 August 1569).

Gilmartin, Kristine: The Thraso-Gnatho subplot in Terence's "Eunuchus". In: *Classical world* 69 (1975/76), pp. 263–267.

Gioffré, M. and Pericle Di Pietro: Il contributo di tre grandi modenesi (Berengario, Falloppia e Folli) all'otologia. In: *Bollettino della Società medico-chirurgica di Modena* 63 (1963), pp. 815–823.

Gitter, Alfred H.: Eine kurze Geschichte der Hörforschung: II. Renaissance. In: *Laryngo-Rhino-Otologie* 69 (1990), pp. 495–500.

Giunta, Tommaso (ed.): *De balneis omnia quae extant apud Graecos, Latinos, et Arabas, tam medicos quam quoscunque ceterarum artium probatos scriptores.* Venice: Iuntas 1553.

Glabbeek, Francis van and Maurits Biesbrouck: Giovanni Baptista Canani (1515–1579 n. s.) and Andreas Vesalius (1514–1564). In: Robrecht Van Hee (ed.): *In the shadow of Vesalius.* Antwerp/Apeldoorn: Garant 2020, pp. 101–148.

Gliozzi, Giuliano: Brasavola, Antonio, detto Antonio Musa. In: *Dizionario Biografico degli Italiani* 14 (1972), pp. https://www.treccani.it/enciclopedia/brasavola-antonio-detto-antonio-musa_(Dizionario-Biografico).

Gola, G.: *L'orto botanico: quattro secoli di attività (1545–1945).* Padua: Edit. Liviana 1947.

Gryll, Lorenz: *Oratio de peregrinatione studii medicinalis ergo suscepta.* [Ingolstadt] 1566.

Gurlt, E.: *Geschichte der Chirurgie und ihrer Ausübung. Volkschirurgie – Altertum – Mittelalter – Renaissance.* Vol. 2. Berlin: Hirschwald 1898.

Gysel, C.: Gabriele Falloppio: His school and the evolution of dentofacial morphology in the 16th century In: *L'orthodontie française* 44 (1973), pp. 313–342.

Haller, Albrecht von: *Bibliotheca botanica.* Vol. 1. London: apud Carol. Heydinger 1771.

Haller, Albrecht von: *Bibliotheca anatomica.* Vol. 1. Zürich: Orell, Gessner et Fuessli 1774.

Haller, Albrecht von: *Bibliotheca chirurgica.* Vol. 1. Basel: apud Joh. Schweighauser 1774.

Hausmann, Frank-Rutger: *Bibliographie der deutschen Übersetzungen aus dem Italienischen von den Anfängen bis zur Gegenwart.* Vol. 1: *Von den Anfängen bis 1730.* Tübingen: Max Niemeyer Verlag 1992 (reprint 2018).

Heinemann, Franz: Die Henker und Scharfrichter als Volks- und Viehärzte seit Ausgang des Mittelalters. In: *Schweizerisches Archiv für Volkskunde* 4 (1900), pp. 1–16.

Helbig, C. E.: Zu dem Schrifttume über den Condom. In: *Reichs-Medicinal-Anzeiger* 32 (1907), pp. 405–407 and pp. 424–426.

Helbig, C. E.: Zur Geschichte der mechanischen Vorbeugemittel gegen Schwängerung und geschlechtliche Ansteckung. In: *Anthropophyteia: Jahrbücher für folkloristische Erhebungen und Forschungen zur Entwicklungsgeschichte der geschlechtlichen Moral* 10 (1913), pp. 3–12.

Herrlinger, Robert and Edith Feiner: Why did Vesalius not discover the Fallopian tubes? In: *Medical History* 8 (1964), pp. 335–341.

Herrmann, Sabine: Ein Preuße in Venedig: Der Botaniker Melchior Wieland (ca. 1520–1589), Pionier der botanischen Feldforschung in der Levante. In: *Sudhoffs Archiv* 99 (2015), pp. 1–14.

Herzog, Markwart: Scharfrichterliche Medizin. Zu den Beziehungen zwischen Henker und Arzt, Schafott und Medizin. In: *Medizinhistorisches Journal* 29 (1994), pp. 309–322.

Hessus, Paulus: *Defensio XX. problematum Melchioris Guilandini aduersus quae Petr. Andreas Matthaeolus ex centum scripsit.* Padua: Olmi 1562.

Himes, N. E.: *Medical history of contraception.* New York: Schocken 1970.

Houtzager, H. L. and O. P. Bleker: Anatomie en fysiologie van het foetale hart en de bloedvaten. In: *Geschiednis der geneeskunde* 9 (2002/2003), pp. 294–305.

Hsu, Kuang-Tai: Gabriele Falloppio's 'De medicatis aquis' as a major source of Nicolaus Steno's earliest geological writing "Dissertatio physica de thermis". In: *Philosophy and the History of Science: A Taiwanese Journal* 2 (1993), pp. 77–104.

Hyrtl, Joseph: *Lehrbuch der Anatomie des Menschen mit Rücksicht auf physiologische Begründung und praktische Anwendung.* Prague: Ehrlich 1846.

Imperatori, Charles J.: Gabriele Fallopius, circa 1523–1562. His contributions to the anatomy of the nose, throat and ear, and his invention and use of the nasal snare. In: *Laryngoscope* 58 (1948), pp. 431–447.

King, Helen: *The one-sex body on trial. The classical and early modern evidence.* Farnham: Ashgate 2013.

Kinzelbach, Annemarie: Erudite and honoured artisans? Performers of body care and surgery in early modern German towns. In: *Social History of Medicine* 27 (2014), pp. 668–688.

Klestinec, Cynthia: Medical education in Padua. Students, faculty and facilities. In: Ole Peter Grell, Andrew Cunningham, and Jon Arrizabalaga (eds.): *Centres of medical excellence? Medical travel and education in Europe, 1500–1789.* Abingdon/New York: Ashgate 2010, pp. 193–220.

Klestinec, Cynthia: Practical experience in anatomy. In: Charles T. Wolfe and Ofer Gal (eds.): *The body as object and instrument of knowledge. Embodied empiricism in early modern science.* Dordrecht: Springer 2010, pp. 33–57.

Klestinec, Cynthia: *Theaters of anatomy. Students, teachers, and traditions of dissection in Renaissance Venice.* Baltimore, MD: Johns Hopkins University Press 2011.

Klestinec, Cynthia: Theater der Anatomie. Visuelle, taktile und konzeptuelle Lernmethoden. In: Ludger Schwarte and Jan Lazardzig (eds.): *Spuren der Avantgarde: Theatrum anatomicum: Frühe Neuzeit und Moderne im Kulturvergleich.* Berlin/New York: De Gruyter 2011, pp. 75–96.

Klestinec, Cynthia: Renaissance surgeons. Anatomy, manual skill and the visual arts. In: Peter M. Distelzweig, Benjamin Goldberg, and Evan R. Ragland (eds.): *Early modern medicine and natural philosophy.* Dordrecht: Springer 2016, pp. 43–58.

Klestinec, Cynthia: Vesalius amont the surgeons. In: *Journal of Medieval and Early Modern Studies* 48 (2018), pp. 125–151.

Kornell, Monique: Illustrationes from the Wellcome Library: Vesalius' method of articulating the skeleton and a drawing in the collection of the Wellcome Library. In: *Medical History* 44 (2000), pp. 97–110.

Kothary, Piyush C. and Sarla P. Kothary: Gabriele Fallopio. In: *International Surgery* 60 (1975), pp. 80–81.

Kramer, Matthias: *Das herrlich-große Teutsch-Italiänische Dictionarium oder Wort- und Red-Arten-Schatz.* Part 2. Nürnberg: In Verlegung Johann Andreä Endters Seel. Söhnen 1702.

Kristeller, Paul Oskar: *Iter italicum* V. London: Warburg Institute 1990.

Landi, Bassiano: *De humana historia, vel singularum hominis partium cognitione libri duo.* Basel: per Ioannem Oporinum 1542.

Laqueur, Thomas W.: *Making sex. Body and gender from the Greeks to Freud.* Cambridge, MA/London: Harvard University Press 1990.

Leclerc, H[enri]: Un naturaliste irascible. P. A. Matthiole de Sienne. In: *Janus* 31 (1927), pp. 334–345.

Macchelli, Niccolò: *Tractatus de morbo gallico scriptus in gratiam iuniorum medicorum almi collegii mutinensi.* Venice: apud Andream Arrivabenum 1555.

Macchelli, Niccolò: *Razae libellus de peste de graeco in latinum sermonem versus per Nicolaum Macchellum.* Venice: apud Andream Arrivabenum 1555.

Macchi, Veronica, Andrea Porzionato, Aldo Morra, and Raffaele Caro: Gabriel Falloppius (1523–1562) and the facial canal. In: *Clinical Anatomy* 27 (2014), pp. 4–9.

Mache, Ursula: *Anatomischer Unterricht in Padua im 16. Jahrhundert. Edition, Übersetzung und Kommentierung der Aufzeichnung eines böhmischen Studenten.* Diss. med. dent. Regensburg. Duisburg: WiKu-Verlag 2019.

MacLean, Ian: The interpretation of natural signs. Cardano's *De subtilitate* versus Scaliger's *Exercitationes.* In: Brian Vickers (ed.): *Occult & scientific mentalities in the Renaissance.* Cambridge: Cambridge University Press 1986, pp. 231–252.

Maclean, Ian: *Logic, signs and nature in the Renaissance. The case of learned medicine.* Cambridge: Cambridge University Press 2002.

MacLean, Ian: Trois facultés de médecine au XVIe siècle: Padoue, Bâle, Montpellier. In: Madeleine Fragonard and Michel Bideau (eds.): *Les échanges entre les universités européennes à la Renaissance.* Geneva: Droz 2004, pp. 349–358.

MacLean, Ian: Diagrams in the defence of Galen. Medical uses of tables, squares, dichotomies, wheels, and latitudes, 1480–1574. In: Sachiko Kusukawa and Ian MacLean (eds.): *Transmitting knowledge. Words, images, and instruments in early modern Europe.* Oxford: Oxford University Press 2006, pp. 135–164.

Makatsariya, Nataliya Aleksandrovna: Габриэлъ Фаллопий. In: *Obstetrics, Gynecology and Reproduction* 10 (2016), pp. 123–124.

Mandressi, Rafael: *Le regard de l'anatomiste. Dissections et invention du corps en Occident.* Paris: Seuil 2003.

Maranta, Bartolomeo: *Methodi cognoscendorum simplicium libri tres.* Venice: Ex officina Erasmiana Vincentiij Valgrisij 1559.

Marchetti, V. and G. Patrizi: Castelvetro, Ludovico. In: *Dizionario Biografico degli Italiani* 22 (1979). https://www.treccani.it/enciclopedia/antonio-fracanzani_ (Dizionario-Biografico)/.

Martin, Craig: *Renaissance meteorology: Pomponazzi to Descartes.* Baltimore, MD: Johns Hopkins University Press 2011.

Martinotti, Giovanni: *L'insegnamento dell' anatomia in Bologna prima del secolo XIX.* Bologna: Azzoguidi 1911.

Massa, Niccolò: *Liber introductorius anatomiae, sive dissectionis corporis humani.* Venice: In aedibus Francisci Bindoni ac Maphei Pasini 1536.

Mattioli, Pietro Andrea: *Commentarii in libros sex Pedacii Dioscoridis Anazarbei de medica materia.* Venice: apud Vincentium Valgrisium 1554.

Mattioli, Pietro Andrea: *Commentarii secundo aucti in libros sex Pedacii Dioscoridis Anazarbei de medica materia.* Venice: apud Vincentium Valgrisium 1558.

Mattioli, Pietro Andrea: *Epistola de bulbocastaneo, oloconitide, mamira, traso, moly, doronico, grano zelin, zedoaria, zurumbeto, carpesio.* Prague: Cantor 1558.

Mattioli, Pierandrea: *Epistolarum medicinalium libri quinque.* Prague: Valgrisius 1561.

Mattioli, Pierandrea: *Adversus XX. problemata Melchioris Guilandini disputatio ad Iohannem Cratonem.* In: Hessus, Defensio (1562), pp. 133–151.

Mazzuchelli, Giammaria: *Gli scrittori d'Italia cioè notizie storiche, e critiche intorno alle vite, e agli scritti dei letterati italiani.* Vol. 1, 1 and Vol. 2, 3. Brescia: Bossini 1773 and 1762.

McVaugh, Michael R.: *The rational surgery of the Middle Ages.* Tavarnuzze/Impruneta: SISMEL 2006.

McVaugh, Michael R.: When universities first encountered surgery. In: *Journal of the History of Medicine and Allied Sciences* 72 (2016), pp. 6–20.

Mercuriale, Girolamo: De decoratione liber [...] ex Hieronymi Mercurialis [...] explicationibus. A Iulio Mancino exceptus & in capita redactus. In: idem: *De morbis cutaneis et de omnibus corporis humani excrementis tractatus*. Venice: apud Paulum Meietum 1585 (separate pagination).

Mercuriale, Girolamo: *De decoratione liber [...] ex Hieronymi Mercurialis [...] explicationibus. A Iulio Mancino exceptus & in capita redactus.* Frankfurt: apud Ioannem Wechelum 1587.

Mercuriale, Girolamo: *De decoratione liber [...] ex Hieronymi Mercurialis [...] explicationibus. A Iulio Mancino exceptus & in capita redactus.* Venice: apud Iuntas 1601.

Mercuriale, Girolamo: *De morbis cutaneis*. Transl. by Richard L. Sutton as *Sixteenth century physician and his methods. Mercurialis on diseases of the skin*. Kansas City, MO: Lowell Pr. 1986.

Mieli, Aldo (ed.): *Gli scienziati italiani dall'inizio del medio evo ai nostri giorni*. Vol. 1, part 1. Rome: Nardecchia Editore 1921.

Minadoi, Giovanni Tommaso: *De humani corporis turpitudinibus cognoscendis et curandis libri tres*. Padua: apud Franciscum Bolzettam 1600.

Minelli, Alessandro (ed.): *L'orto botancio di Padova, 1545–1995*. 2nd edn. Venice: Marsilio 1998.

Montalenti, Giuseppe: Gabrielle Falloppia, anatomo e medico. In: Aldo Mieli (ed.): *Gli scienziati italiani*, vol. 2, part 1. Rome: Casa editrice Leonardo da Vinci 1923 [published only in or after 1928], pp. 44–59.

Morgagni, Giovanni Battista: Ad illustrissimum et celeberrimum virum D. Jo. Mariam Lancisium intimum cubicularium, et archiatrum Pontificium, epistola (Padua, 20 October 1713). In: Eustachius, Bartholomaeus: *Tabulae anatomicae*. Rom: ex officina typographica Francisci Gonzagae 1714, pp. XVI–XVIII.

Mortazavi, M. M., N. Adeeb, B. Latif, K. Watanabe, A. Deep, C. J. Griessenauer, R. S. Tubbs, and T. Fukushima: Gabriele Fallopio (1523–1562) and his contributions to the development of medicine and anatomy. In: *Child's Nervous System* 29 (2013), pp. 877–880.

Münster, Ladislao: La laurea di Gabriele Falloppia allo Studio di Ferrara (1552). In: *Ferrara viva* 5 (1965), pp. 181–206.

Münster, Ladislao: Die Universität zu Ferrara und die Blütezeit ihrer Medizinschule im 15. und 16. Jahrhundert. In: *Die Grünenthal Waage* 7 (1968), pp. 49–60.

Muratori, Giulio: The academic career and anatomical teaching of G. B. Cananus at St. Dominic and the anatomical theatres of the University of Arts and Medicine of Ferrara. In: *Acta anatomica* Suppl. 56 (ad vol. 73) (1969), pp. 308–324.

Muratori, Giulio E. and Cesare Menini: Contributi allo studio della storia dell'anatomia e della medicina nell'ateneo ferrarese nel 1500. In: *Annali Università Ferrara* 5 (1946), pp. 19–103.

Muratori, G[iulio] and D[elfino] Bighi: Andrea Vesalio, G. B. Canano e la rivoluzione rinascimentale dell'anatomia e della medicina. In: *Acta medicae historiae patavina* 10 (1963–1964), pp. 51–95.

Natale, Gianfranco, Guido Bocci and Domenico Ribatti: Scholars and scientists in the history of the lymphatic system. In: *Journal of Anatomy* (2017). doi: 10.1111/joa.12644.

Nowosadtko, Jutta: Wer Leben nimmt, kann auch Leben geben. Scharfrichter und Wasenmeister als Heilkundige in der Frühen Neuzeit. In: *Medizin, Gesellschaft und Geschichte* 12 (1993), pp. 43–74.

Nunn, Hillary M.: *Staging anatomies: Dissection and spectacle in early Stuart tragedy.* Aldershot: Ashgate 2005.

Nutton, Vivian: Humanist surgery. In: Andrew Wear, Roger French, and Iain Lonie (eds.): *The medical renaissance of the sixteenth century.* Cambridge: Cambridge University Press 1985, pp. 75–99.

Nutton, Vivian: Medical humanism, a problematic formulation? In: Arts et savoir 15 (2021) (= Revisiting humanism in Renaissance Europe), online edition https:// doi.org/10.4000/aes.3925.

Nutton, Vivian: The rise of medical humanism: Ferrara, 1464–1555. In: *Renaissance Studies* 11 (1997), pp. 2–19.

Olivieri, Achille: Cavalli, Marino. In: *Dizionario biografico degli italiani, vol. 22.* https://www.treccani.it/enciclopedia/marino-cavalli_(Dizionario-Biografico)/ 1979, pp.

Olivieri, Achille: Experimentum e structura: da Falloppia a Harvey. In: Giuseppe Ongaro, Maurizio Rippa Bonati and Gaetano Thiene (eds.): *Harvey e Padova.* Treviso: Antilia 2006, pp. 175–194.

O'Malley, Charles D.: Gabriele Falloppia's account of the cranial nerves. In: Gernot Rath and Heinrich Schipperges (eds.): *Medizingeschichte im Spektrum. Festschrift zum fünfundsechzigsten Geburtstag von Johannes Steudel.* Wiesbaden: Franz Steiner Verlag 1961, pp. 132–137.

O'Malley, Charles D. Gabriele Falloppia (c. 1523–1562). In: *Investigative Urology* 4 (1966), pp. 95–96.

O'Malley, Charles D.: Gabriele Falloppia's account of the orbital muscles. In: Lloyd G. Stevenson and Robert P. Multhauf (eds.): *Medicine, science and culture. Historical essays in honor of Owsei Temkin.* Baltimore, MD: Johns Hopkins University Press 1968, pp. 77–85.

O'Malley, Charles D.: Medical education during the Renaissance. In: idem (ed.): *History of medical education.* Berkeley: University of California Press 1970, pp. 89–102.

O'Malley, Charles D.: Falloppio, Gabriele. In: *Complete dictionary of scientific biography.* Vol. 4. New York: Charles Scribner's Sons 2008, pp. 519–521.

O'Malley, Charles D. and Edwin Clarke: The discovery of the auditory ossicles. In: *Bulletin of the History of Medicine* 35 (1961), pp. 419–441.

Öncel, Çağatay: One of the great pioneers of anatomy: Gabriele Falloppio (1523–1562). In: *Bezmialem Science* 3 (2016), S. 123–126.

Ongaro, Giuseppe: La scoperta della circolazione polmonare e la diffusione della *Christianismi restitutio* di Michele Serveto nel XVI secolo in Italia e nel Veneto. In: *Episteme. Rivista critica di storia delle scienze mediche e biologiche* 1 (1971), pp. 3–44.

Ongaro, Giuseppe: Bassiano Landi e Andrea Vesalio. In: *Atti e memorie dell'Accademia patavina di scienze, lettere ed arti. Memorie della classe di scienze matematiche e naturali* 110, part 2 (1998), pp. 33–54.

Ongaro, Giuseppe: Falloppia, Gabriel (aka Falloppio, Gabriel; Fallopius). In: William F. Bynum and Helen Bynum (eds.): *Dictionary of medical biography.* Vol. 2: C–G. Westport, CT/London: Greenwood Press 2007, pp. 473–474.

Oxford dictionary of the Renaissance. Ed. by Gordon Campbell. Oxford/New York: Oxford University Press 2003.

Palmer, Richard: *The Studio of Venice and its graduates in the sixteenth century.* Padua: Edizioni Lint 1983.

Palmer, Richard: Falloppia. In: Roy Porter (ed.): *Dizionario biografico della storia della medicina e delle scienze naturali.* Vol. 2. Milan: Franco Maria Ricci 1987, pp. 13–14.

Palmer, Richard: "In this our lightye and learned tyme". Italian baths in the era of the Renaissance. In: Roy Porter (ed.): *The medical history of waters and spas.* London: Wellcome Institute 1990, pp. 14–22.

Panetto, Monica and Vito Terribile Wiel Marin: Gabriele Falloppia (1523–1562): l'experientia tra anatomia e Riforma. Con nuovi documenti relativi alla ricognizione del 1996. In: *Studi storici Luigi Simeoni* 51 (2001), pp. 272–306.

Parent, André: Berengario da Carpi and the renaissance of brain anatomy. In: *Frontiers in Neuroanatomy* 13 (2019). doi: 10.3389/fnana.2019.00011 (special issue on "History of neuroscience", pp. 1–14).

Pardi, Giuseppe: *Titoli dottorali conferiti dallo studio di Ferrara nei sec. XV e XVI* (repr. Bologna: Forni 1970). Lucca: Marchi 1901.

Park, Katherine: Natural particulars: Medical epistemology, practice, and the literature of healing springs. In: Anthony Grafton and Nancy Siraisi (eds.): *Natural particulars: nature and the disciplines in Renaissance Europe.* Cambridge, MA/London: MIT-Press 1999, pp. 347–367.

Park, Katharine and Robert A. Nye: Destiny is anatomy. In: *The New Republic* 18 February 1991, pp. 53–57.

Patricius, Andreas: Dedicatory epistle to Gabriele Falloppia, 5 January 1558. In: Melchior Wieland and Conrad Gessner: *De stirpibus aliquot epistolae.* Padua: apud Gratiosum Perchacinum 1558, foll. 2r–4v.

Petit, Caroline: Gadaldini's library. In: *Mnemosyne* 60 (2007), pp. 132–138.

Petrucci, Giuseppe: *Vite e ritratti di XXX. illustri Ferraresi.* Bologna: Litografia Zannoli 1833.

Phillips, Kim M.: The breasts of virgins: Sexual reputation and young women's bodies in medieval culture and society. In: *Cultural and Social History* 15 (2018), pp. 1–19. doi: 10.1080/14780038.2018.1427341.

Politzer, Adam: *Geschichte der Ohrenheilkunde.* Part 1. *Von den ersten Anfängen bis zur Mitte des 19. Jahrhunderts.* Stuttgart: Enke 1907.

Preti, Cesare: Guidi, Guido. In: *Dizionario Biografico degli Italiani* 61 (2004). https://www.treccani.it/enciclopedia/guido-guidi_res-65f9fb9e-87ee-11dc-8e9d-0016357eee51_(Dizionario-Biografico)/.

Ragland, Evan R.: "Making trials" in sixteenth- and early seventeenth-century European academic medicine. In: *Isis* 108 (2017), pp. 503–528.

Raimondi, Carlo: Una lettera inedita di P.A. Mattioli a Gabriele Falloppio. In: *Bullettino senese di storia patria* 10 (1903), pp. 286–289.

Raimondi, Carlo: Lettere di P.A. Mattioli ad Ulisse Aldrovandi. In: *Bullettino senese di storia patria* 13 (1906), pp. 121–185.

Rankin, Alisha: *The poison trials. Wonder drugs, experiment, and the battle for authority in Renaissance science.* Chicago, IL: Chicago University Press 2021.

Rippa Bonati, Maurizio: Su un insegnamento di anatomia tenuto da Bassiano Landi. In: *Atti e memorie dell'Accademia patavina di scienze, lettere ed arti. Memorie della classe di scienze matematiche e naturali* 110, part 2 (1998), pp. 55–61.

Rippa Bonati, Maurizio: Girolamo Fabrici d'Acquapendente: per una bio-crono-bibliografia. In: idem and Pardo-Tomàs, *Il teatro dei corpi* (2004), pp. 268–277.

Rippa Bonati, Maurizio and José Pardo-Tomàs (eds.): *Il teatro dei corpi. Le pitture colorate d'anatomia di Girolamo Fabrici d'Acquapendente.* Milan: Mediamed 2004.

Robison, Kira: *Healers in the making: Students, physicians, and medical education in medieval Bologna (1250–1550).* Leiden/Boston, MA: Brill 2020.

Robortello, Francesco: *De convenientia supputationis Livianae Ann. cum marmoribus Rom. quae in capitolio sunt. De arte, sive ratione corrigendi veteres authores, disputatio. Emendationum libri duo.* Padua: apud Innocentium Olmum 1557.

Romagnoli, Giovanni: Una lettera importante del famoso anatomo Gabriele Falloppia nella "Collezione autografi Piancastelli" della Biblioteca comunale di Forlì con alcune considerazioni sulla sua laurea dottorale e sulla sua malattia mortale. In: *Giornale di batteriologia, virologia, ed immunologia ed annali dell'Ospedale Maria Vittoria di Torino* 60 (1967), pp. 345–362.

Rondelet, Guillaume: *Libri de piscibus marinis in quibus verae piscium effigies expressae sunt.* Lyon: Bonhomme 1554.

Rossettino, Giovanni: Canzone nella morte del Falloppia. In: idem: Catalogo sopra li dottori, che leggono nel [sic] studio di Padova. Padua: Nella Stamparia del Griffio 1763 (no pagination).

Rossi, Francesco: Medaglie. In: Mina Gregori (ed.): *I Campi. Cultura artistica cremonese del Cinquecento.* Milan: Museo Civico 1985, pp. 347–368.

Roth, Moritz: *Andreas Vesalius Bruxellensis.* Berlin: Reimer 1892.

Rudwick, Martin J. S.: *The meaning of fossils. Episodes in the history of paleontology.* London/New York: Macdonald/American Elsevier 1972.

Sacchi, Rossana: Intorno agli Anguissola. In: *Sofonisba Anguissola e le sue sorelle. Catalogo della mostra.* Milan: Leonardo Arte 1994, pp. 345–360.

Saliceto, Guilelmus de: De decoratione. In: idem: *Summa conservationis et curationis.* Venice: Octavianus Scotus 1502, foll. 96v–105r.

Savonarola, Giovanni Michele: *De balneis et thermis naturalibus omnibus Italiae. Dedicated to Borso d'Este.* Ferrara: Andreas Belfortis 1485.

Scarpa, Antonio: *Elogio storico di Giambatista Carcano Leone, professore di notomia nella Università di Pavia.* Milan: Stamperia reale 1813.

Schieß, Traugott: *Briefe aus der Fremde von einem Zürcher Studenten der Medizin (Dr. Georg Keller) 1550–1558 (= Neujahrblatt Nr. 262).* Zürich: Kommissionsverlag von Fäsi & Beer 1906.

Scipio, Rosario (ed.): *Girolamo Fabrici l'Acquapendente.* Viterbo: Agnesotti 1978.

Scott, John Russell: *Catalogue of the manuscripts remaining in Marsh's Library, Dublin.* Dublin: A. Thom and Co. 1913.

Seidel, Bruno: *Liber morborum incurabilium causas, mira brevitate, summa lectionis jucuditate exhibens.* Frankfurt: apud Joannem Wechelum 1593.

Senfelder, Leopold: Georg Handsch von Limus. Lebensbild eines Arztes aus dem XVI. Jahrhundert. In: *Wiener klinische Rundschau* (1901), pp. 495–499, 514–516 and 533–535.

Sigonio, Carlo: *Emendationum libri duo.* Venice: Aldus 1557.

Siraisi, Nancy: Vesalius and the reading of Galen's teleology. In: *Renaissance Quarterly* 50 (1997), pp. 1–37.

Siraisi, Nancy G.: *Historia, actio, utilitas*: Fabrici e le scienze della vita nel Cinquecento. In: Rippa Bonati/ Pardo-Tomàs, *Il teatro dei corpi* (2004), pp. 63–73.

Smolka, Josef and Marta Vaculínová: Renesanční lékař Georg Handsch (1529–1578). In: *DVT – Dějiny věd a techniky* 43 (2010), pp. 1–26.

Speert, Harold: Obstetric-gynecologic eponyms: Gabriele Falloppio and the fallo-pian tubes. In: *Obstetrics & Gynecolog* 6 (1955), pp. 467–470.

Steno, Nicolaus: *Disputatio physica de thermis.* Amsterdam: apud Joannem Ravest-einium 1660.

Steno, Nicolaus: De thermis. In: Gustav Scherz (ed.): *Steno: Geological papers.* Odense: Odense University Press 1969, pp. 45–63.

Steudel, Johannes: Die Entdeckung der Venenklappen. In: *Deutsche medizinische Wochenschrift* 80 (1955), pp. 1913–1915.

Stolberg, Michael: *A history of palliative care 1500–1970. Concepts, practices and ethical challenges.* Cham: Springer 2017.

Stolberg, Michael: A woman down to her bones. The anatomy of sexual difference in the sixteenth and early seventeenth centuries. In: *Isis* 94 (2003), pp. 274–299.

Stolberg, Michael: 'Abhorreas pinguedinem': Fat and obesity in early modern med-icine (c. 1500–1750). In: *Studies in the History and Philosophy of Biology and Bio-medical Sciences* 43 (2012), pp. 370–378.

Stolberg, Michael: Bedside teaching and the acquisition of practical skills in mid-sixteenth-century Padua. In: *Journal of the History of Medicine and Allied Sciences* 69 (2014), pp. 633–661.

Stolberg, Michael: „Cura palliativa". Begriff und Diskussion der palliativen Krank-heitsbehandlung in der vormodernen Medizin (ca. 1500–1850). In: *Medizinhistor-isches Journal* 42 (2007), pp. 7–29.

Stolberg, Michael: Empiricism in sixteenth-century medical practice. The note-books of Georg Handsch. In: *Early Science and Medicine* 18 (2013), pp. 487–516.

Stolberg, Michael: Learned physicians and everyday medical practice in the Renais-sance. Berlin: De Gruyter 2022.

Stolberg, Michael: Learning anatomy in late sixteenth-century Padua. In: *History of Science* 56 (2018), pp. 381–402.

Stolberg, Michael: The doctor-patient relationship in the Renaissance. In: *European Journal for the History of Medicine and Health* 1 (2021), pp. 1–29.

Stolberg, Michael: Teaching anatomy in Post-Vesalian Padua. An analysis of stu-dent notes. In: *Journal of Medieval and Early Modern Studies* 48 (2018), pp. 61–78.

Stolberg, Michael: Tödliche Menschenversuche im 16. Jahrhundert. In: *Deutsches Ärzteblatt*, A, 111 (2014), pp. 2060–2062.

Stolberg, Michael: Training future practitioners. Medical education in sixteenth- and early-seventeenth-century Padua and Montpellier from the students' per-spective. In: Delia Gavrus and Susan Lamb (eds.): *Medical* education. *Historical case studies of teaching, learning, and belonging in medicine in honour of Jacalyn Duffin.* Montreal/Kingston: McGill-Queen's University Press 2022, pp. 112–135.

Storchová, Lucie: Handsch, Georg. In: eadem (ed.): *Companion to central and eastern European humanism.* Vol. 2: *Czech lands.* Part 1. Berlin: De Gruyter 2020, pp. 512–522.

Tagault, Jean: *De chirurgica institutione libri 5.* Paris: Wechelius 1543.

Tagliacozzi, Gaspare: *De curtorum chirurgia per insitionem libri duo.* Venice: apud Robertum Meiettum 1597.

Temkin, Owsei: *The falling sickness. A history of epilepsy from the Greeks to the be-ginnings of modern neurology.* 2nd edn. Baltimore, MD/London: Johns Hopkins University Press 1971.

Temkin, Owsei: *Galenism: Rise and decline of a medical philosophy.* Ithaca, NY: Cor-nell University Press 1973.

Theophrastos: *De historia plantarum libri IX.* Paris: Christian Wechel 1529.

Theophrastos: Περὶ φυτῶν ἱστορία. Venice: Turrisanus 1529.

Thorndike, Lynn: *A history of magic and experimental science*. Vol. 5. New York: Columbia University Press 1941.

Tiraboschi, Girolamo: *Biblioteca modenese o notizie della vita e delle opere degli scrittori natii degli stati del serenissimo signor duca di Modena*. Vol. 2. Modena: Società Tipografica 1782.

Tomasini, Giacomo Filippo: *Bibliothecae patavinae manuscriptae publicae et privatae*. Udine: Typis Nicolai Schiratti 1639.

Tomasini, Giacomo Filippo: *Gymnasium patavinum [...] libris V. comprehensum*. Utini: Ex typographia Nicolai Schiratti 1654.

Tomasini, Giacomo Filippo: *Illustrivm virorum elogia iconibus exornata*. Padua: apud Donatum Pasquardum & Socium 1630.

Trincavella, Vittore: *Consilia medica post editionem venetam et lugdunensem, accessione CXXVIII consiliorum locupletata, et per locos communes digesta*. Basel: apud Conradum Valdkirchium 1587.

Tsihuaka, Barbara I.: Falloppia, Gabriele. In: Werner E. Gerabek, Bernhard D. Haage, Gundolf Keil, and Wolfgang Wegner (eds.): *Enzyklopädie Medizingeschichte*. Berlin and New York: De Gruyter 2005, pp. 391–392.

Van Gijn, Jan and Joost P. Gijselhart: Falloppius and his uterine tubes. In: *Nederlands tijdschrift voor geneeskunde* A 155 (2011), pp. 3639.

Van Hee, Robrecht H.: The relationship between Vesalius and the Borgarucci family. In: *Acta chirurgica belgica* 117 (2017), pp. 329–343.

Vedriani, Lodovico: *Dottori modonesi di teologia, filosofia, legge canonica, e civile con i suoi ritratti dal naturale in rame. Et altri letterati insigni per l'opere, e dignità loro*. Modena: Per Andrea Cassiani 1665.

Vesalius, Andreas: *Anatomicarum Gabrielis Falloppii observationum examen*. Venice: de Franciscis 1564.

Vesalius, Andreas: *Anatomicarum Gabrielis Fallopii observationum examen. Magni humani corporis fabricae operis appendix*. Ed. by Johann Jessen. Hannover: Typis Wechelianis apud Claudium Marnium et heredes Ioan. Aubrii 1609.

Vesalius, Andreas: *De humani corporis fabrica libri septem*. Basel: Oporinus 1543.

Visentin, Michele: *Falloppio & Guilandino. Una "liaison dangereuse" nella Padova del Cinquecento* (https://www.culturagay.it/saggio/482, accessed 28 December 2021).

Wells, Walter: Gabriel Fallopio: One of the sixteenth century founders of modern anatomy. Also a distinguished physician and surgeon 1523–1562. In: *Laryngoscope* 58 (1948), pp. 33–42.

Welsch, Georg Hieronymus: Consiliorum medicinalium centuriae quatuor. In: idem: *Curationum exotericarum chiliades II et consiliorum medicinalium centuriae IV*. Ulm: Kühn 1676 (separate pagination).

Wieland, Melchior: Melchior Guilandinus borussus Conrado Gesnero medico S. P.D. In: idem and Conrad Gessner: *De stirpium aliquot nominibus vetustis ac novis, quae multis iam seculis vel ignorarunt medici, vel de eis dubitarunt*. Basel: apud Episcopium iuniorem 1557, pp. 7–23.

Wieland, Melchior: *Apologiae adversus Petr. Andream Matheolum liber primus qui inscribitur Theon*. Padua: apud Gratiosum Perchacinum 1558.

Wieland, Melchior: *Papyrus, hoc est commentarius in tria C. Plinj maioris de papyro capita*. Venice: apud M. Antonium Ulmum 1572.

Wierus, Ioannis: *Medicarum observationum rararum liber I*. Basel: Io. Oporinus 1567.

Wightman, William P.D.: Quid sit methodus? "Method" in the sixteenth century medical teaching and "discovery". In: *Journal of the History of Medicine* 19 (1964), pp. 360–376.

Youseff, Mohamed H.: The history of the condom. In: *Journal of the Royal Society of Medicine* 86 (1993), pp. 226–228.

Zaffarini, Niccolò: *Scoperte anatomiche di Gio. Bat. Canani medico e chirurgo Ferrarese del secolo XVI. Illusttrate, ed arrichite di diverse notizie spettanti alla vita di lui e d'altri celebri professori ed anatomici ferraresi*. Ferrara: Per Gaetano Bresciani 1909.

Zamoyski, Jan: *Oratio, habita Patavii, in funere excellentissimi viri, Gabrielis Falloppii, IIII. Id. Oct. MDLXII*. Padua: Ulmus 1562.

Zampieri, Fabio, Gaetano Thiene, Christina Basso und Alberto Zanatta: The three fetal shunts. A story of wrong eponyms. In: *Journal of Anatomy* 238 (2021), pp. 1028–1035.

Zanier, Giancarlo: *Medicina e filosofia tra '500 e '600*. Milan: Angelo 1983.

Zaun, Stefanie, Daniela Watzke, and Jörn Steigerwald (eds.): *Imagination und Sexualität. Pathologien der Einbildungskraft im medizinischen Diskurs der frühen Neuzeit*. Frankfurt: Klostermann 2004.

Notes

1 Digital image under https://doi.org/10.7891/e-manuscripta-4453.
2 Ed. by Campori, Lettere (1864), pp. 439–441; reprint of Campori's quite faulty transcription in Di Pietro, Epistolario (1970), pp. 73–75.
3 Ed. in Di Pietro, Epistolario (1970).
4 Ed. in Raimondi, Lettere (1906).
5 Ed. in De Toni, Spigolature (1911).
6 Ed. in Di Pietro, Epistolario (1970), pp. 66–67.
7 Ed. ibid., pp. 67–68.
8 Title according to Scott, Catalogue (1913), p. 85. I have not seen this manuscript, which is presumably an English translation of Falloppia's *De morbo gallico*.
9 Ed. in Di Pietro, Epistolario (1970), p. 21.
10 Photographical reproduction in Imperatori, Fallopius (1948), p. 434; ed. in Di Pietro, Epistolario (1970), p. 22.
11 Ed. in Di Pietro, Epistolario (1970), pp. 69–71.
12 Ed. in Raimondi, Lettere (1906), pp. 165–166.
13 The notes end in March, at the beginning of the parts on the French disease because the unidentified scribe, as he noted in conclusion, left Padua, having received his doctorate. This is the manuscript Favaro (Favaro, Gabrielle Falloppia (1928), note on p. 7) could not consult and mentioned only as having been sold by the Florentine antique books dealer Lier in 1925, catalogue number 99; according to the accession book of the Wellcome Library (WA/HMM/Li/Acc/4, Wellcome Library Accession Book, Vol. 4, pp. 200–201, accession number 43968), it was bought, in fact, from Lier, in 1925, and the page on n° 99 from Lier's catalogue is still glued in.
14 Ed. in Di Pietro, Epistolario (1970), pp. 28–29; copies in Biblioteca Estense, Modena, Mss. Ital. n. 854 and in British Library, Egerton MS 44, fol. 27.
15 Ed. in Di Pietro, Epistolario (1970), pp. 50–54 and p. 75.
16 Ed. in Di Pietro, Epistolario (1970), p. 73, p. 75 and pp. 76–77.

17 Ed. ibid., p. 50.
18 MS Canon Misc. 115 and 119 clearly were once part of a more voluminous manuscript.
19 Ed. in Favaro, Gabrielle Falloppia (1928), pp. 223–225.
20 Ed. in Di Pietro, Epistolario (1970), pp. 26–27.
21 Ed. ibid., p. 76.
22 Ed. ibid., p. 77.
23 My thanks to Richard Šípek of the Národní Muzeum in Prague for pointing out that letter.
24 According to Kristeller, Iter italicum V (1990), p. 182.
25 Ed. in Di Pietro, Epistolario (1970), pp. 29–30.
26 Ed. ibid., pp. 22–24.
27 According to the entry on Ms. 169 in the online-catalogue of Venetian manuscripts, Nuova Biblioteca Manoscritta.
28 Content: foll. 1v–130v, notes on about 35 *collegia* in Padua (the index on fol. 1r indicates that notes on further 11 collegia (nrs. X to XX) have gone missing; fol. 133r–v, *Pondera graeca*; fol. 134r, *Mensurae medicae*; foll. 134v–139r, *De causticis*; foll. 140r–249r, *De tumoribus praeter naturam* (13 November 1555–16 July 1556); foll. 250r–317v, *De ulceribus*, 1555; foll. 317–367v, *De morbo gallico*; foll. 371r–402v, *De vulneribus*, January 1555, with an index (foll. 401v–402v); foll. 405r–453v, *De balneis* (ending on 9 July 1556); foll. 371r–402v are in a different handwriting and according to a note in the margin (fol. 371r) were given to the writer by Ortolf Marolt.
29 Published by the same printers, the work carries a different title but was published as a second edition.
30 According to the Catalogo del Servizio Bibliotecario Nazionale.
31 According to the Catalogo del Servizio Bibliotecario Nazionale.
32 No later edition known but the various editions of *De tumoribus praeter naturam* contain a short chapter on the *bubo pestilentialis*.
33 The third page is numbered 1.
34 The frontispiece differs slightly from the other edition Avanzi published in 1564 and also indicates the printer ("Ex officina Stellae, Iordanis Ziletti").
35 In the various editions of the *Opera*, the letter ist dated 1 November 1557.

Index

For Product Safety Concerns and Information please contact our EU
representative GPSR@taylorandfrancis.com
Taylor & Francis Verlag GmbH, Kaufingerstraße 24, 80331 München, Germany

9 7 8 1 0 3 2 1 4 9 7 1 4